T0418932

Environment and Innovation
Strategies to Promote Growth and Sustainability

Editors

Clara Inés Pardo Martínez

School of Administration
Universidad del Rosario
Bogotá, Colombia

Alexander Cotte Poveda

Faculty of Economy
Universidad Santo Tomas
Bogotá, Colombia

CRC Press is an imprint of the
Taylor & Francis Group, an **informa** business
A SCIENCE PUBLISHERS BOOK

Cover credit: Front cover photos by Alexander Cotte Poveda.

First edition published 2021
by CRC Press
6000 Broken Sound Parkway NW, Suite 300, Boca Raton, FL 33487-2742

and by CRC Press
2 Park Square, Milton Park, Abingdon, Oxon, OX14 4RN

Library of Congress Cataloging-in-Publication Data

Names: Pardo Martínez, Clara Inés, editor. | Cotte Poveda, Alexander, editor.
Title: Environment and innovation : strategies to promote growth and sustainability / editors, Clara Inés Pardo Martínez, Alexander Cotte Poveda.
Description: First edition. | Boca Raton, FL : CRC Press, [2021] | Includes bibliographical references and index.
Identifiers: LCCN 2021009933 | ISBN 9780367682750 (hardcover)
Subjects: LCSH: Sustainable development. | Technological innovations--Environmental aspects. | Climatic changes--Economic aspects.
Classification: LCC HC79.E5 E5687 2021 | DDC 338.9/27--dc23
LC record available at https://lccn.loc.gov/2021009933

ISBN: 978-0-367-68275-0 (hbk)
ISBN: 978-0-367-68276-7 (pbk)
ISBN: 978-1-003-13671-2 (ebk)

Typeset in Times New Roman
by Radiant Productions

To

Juliana

Foreword

As an expert and researcher in topics related to innovation, economic growth, and development, I want to introduce this book that shows key elements to promote innovation and sustainability in companies to achieve green economy and migrate to low carbon economies.

Innovation plays an important role in environmental performance because the new ideas attempt to achieve cleaner process, technologies and energy, green products, use of waste as raw materials, reduce use of nonrenewable materials, among others. However, it is important to promote bottom-up type public policiea as strategies to apply sustainability. Hence, this book analyses and evaluates different perspectives in the relationship between innovation process, its application, and advantages to environment.

Every chapter of this book shows the importance of innovation sustainability, digital marketing to promote green economy, biotechnology to promote cleaner process and energy, and bio-economy to promote rural development. In the services sector, it evaluates infrastructure project and connection with social improvements, how these processes require environmental education to promote innovation, the tax and incentives strategies in the context of circular economy, the challenges of companies to implement green accountability, application of environment and innovation in the aviation sector and tourist, and the importance of route plan to determine successful factors and lessons learned.

All these elements and concepts allow that this book achieves an important contribution to determine that growth and sustainable development depends on innovation and possibilities and plans to promote these processes in the productive sector through the policies instruments and organizational initiatives. The information and notions of this book are an important input to establish actions and programs to promote from different perspectives innovation and environment that allow and guarantee sustainability productive process through the adequate investments in research and development and generation of new knowledge.

Jose Antonio Ocampo

Professor of Professional Practice in International and Public Affairs
School of International and Public Affairs
Columbia University

Preface

Innovation is a key element to determine new solutions to control contamination, generate new clean process, prevent and mitigate effects of pollution, maintain or increase economic growth and development and guarantee welfare and quality of life. For this reason, the measurement of innovation and relationship with sustainability and climate change is important topic especially in emerging economies that require new strategies to promote innovation to control environmental problems and competitiveness.

In recent years, the concept of sustainable competitiveness has emerged as the set of institutions, policies and factors that make a country productive over the longer term while warranting social and environmental sustainability. This definition implies high-quality growth, natural resource management, social equality, human development and well-being, which should be the ultimate objectives of companies, businesses, society and national development where it is important to determine the effects of environmental performance on competitiveness, Pardo and Cotte (2021a).

Achieving competitiveness requires having access to certain resources and skills. Such factors can be divided based on two perspectives. The first focuses on basic (natural resources, climate, unskilled and semi-skilled labour and debt capital) and advanced factors (information and communications infrastructure, highly educated labour such as engineers and computer scientists and university research institutes in sophisticated disciplines). According to the second perspective, basic factors of competitiveness include education, health, infrastructure well-functioning markets and a fourth industrial revolution in terms of business sophistication, innovation and research and development; these factors are considered fundamental to achieving development, growth and competitiveness. According to these perspectives, environmental management and performance are central in that competitive economies are better positioned to adopt greener technologies and processes and transition to a low carbon or "de-carbonized" economy. For this reason, it is important to analyse how different environmental variables may affect competitiveness, especially in countries with higher levels of biodiversity and natural resources, Pardo and Cotte (2021a).

In this context, this book evaluates through empirical analysis the relationship between environment, innovation and strategies to promote growth and sustainability, climate change and innovation in different countries using various techniques and data base that allow to understand how innovation play an important role in sustainability development and economic growth in emerging economies that is characterized by its environmental wealth, biodiversity and vulnerability to climate change, Pardo and Cotte (2021b).

Results of this study suggests the importance of innovation in the generation of clean process and environmentally friendlily goods and services and prevention and mitigation of climate change, higher innovation should achieve lower pollution, CO_2 emissions, and environmental problems, and higher competitiveness and economic growth, where it is important to develop political instruments to promote innovation process with environmental criteria that allow through innovation to generate a sustainable development, responsible use of natural resources and activities to control and prevent climate change.

Findings in every chapter of this book are an important input to design and apply policies that integrate innovation and environment as key element of economic growth, development and sustainability in countries.

References

Pardo, C.I. and Cotte, A. (2021a). The effects of environmental performance on competitiveness and innovation: A stochastic frontier approach, Working Paper.

Pardo, C.I. and Cotte, A. (2021b). (Ed.). Environmental Sustainability and Development in Organizations Challenges and New Strategies. Routledge, Taylor & Francis & CRC Press.

Clara Inés Pardo Martínez

Alexander Cotte Poveda

List of Reviewers

Anida Yupari, Extractive Industries and Sustainable Development, UNCTAD.

Andrés Alberto Barrios, Universidad de los Andes.

Joachim Venus, Potsdam University.

Anjana Pandey, Mnnit Allahabad.

Rosalina Gonzalez, Universidad de la Salle.

Yarleys Pulgarín Osorio, Universidad de la Salle.

Daniel Torralba, Score – Universidad del Rosario.

Arely Paredes, Conacyt, Facultad de Ciencias, UMDI Sisal.

Edesiri Okoro, Delta State University, Abraka.

Varsha Agarwal, Center for Management Studies, Jain (Deemed-to-be University).

He Zhu, Institute of Geographic Sciences and Natural Resources Research, Key, Laboratory of Regional Sustainable Development Modeling, China.

Luis Doña Toledo, Universidad de Granada.

Segura Saul Serna, Woosong University.

Contents

CHAPTER 1

Analysis of the Relationships Among Between Climate Change and Innovation in Colombia

An Empirical Approach

Alexander Cotte Poveda [1,*] and *Clara Inés Pardo Martínez*[2,*]

Introduction

In recent decades, the increase in anthropogenic greenhouse gas (GHG) emissions and contamination problems has transformed the composition of the atmosphere and changed the global climate system, including the amount and pattern of precipitation, meteorological characteristics, health problems, and some human and biological activities worldwide (UNDP 2004, EPA 2018). GHG emissions have increased by an average of 2.2% per year during recent decades, and in recent years, anthropogenic GHG emissions reached 49 ± 4.5 GtCO-eq, while anthropogenic global warming increased by 0.2°C per decade due to past and ongoing emissions. These increases are related to changes in snow cover, ice cover, sea levels, and precipitation disturbances in many regions of the world and to changes in human health, economic activities, and the characteristics of human settlements (IPCC 2014, 2018).

Currently, governments recognize climate change has a meaningful impact on economies. Therefore, innovative solutions are needed that can respond to the changes generated by climate change. It has become necessary to analyse the role of innovation to support the damages occasioned by this phenomenon (UNEP 2018).

Studies on the relationships among climate change and innovation have been made mainly at macro level in the context of developed countries or BRICS (Brazil, Russia, India, China, and South Africa) countries; for example, Santra

[1] Universidad Santo Tomas, Faculty of Economy, Bogotá, Colombia.
[2] Universidad de Rosario, School of Administration, Bogotá, Colombia.
* Corresponding authors: alexandercotte@uniandes.edu.co; cipmusa@yahoo.com

(2017) determined the effects of eco-technology innovations on the reduction of CO_2 emissions and the increase in productivity, especially in BRICs economics. Aldieri et al. (2019) demonstrated the positive relationship between productivity and environmental innovation in Russian regions. Du and Li (2019) studied 71 economies in the world for the period 1992–2012, finding that green technology innovations have effects only in developed economies and that the effects of green technology innovations on productivity are nonsignificant in developing countries because fewer new green technologies or innovations are applied than in developed countries. Morris (2018) evaluated the effect of innovation on productivity across manufacturing and services firms globally, demonstrating that both are positively related, but the manufacturing and service sectors differ, and the process is very complex and heterogeneous. Ranasinghe (2017) studied Canada and US innovation spending, productivity, and firm size differences, where the main results indicated that subsidies for innovation can be more cost effective and can increase innovation and productivity. Szopik-Depczyńska (2018) calculated the innovation level of national economies in Europe using different indicators, determining that eco-innovations are important for progress and to improve the quality of life that has become fundamental for sustainable development. These studies show the importance of connecting innovation with solutions to address issues related with climate change.

With this background, this study seeks to analyse the relationships among innovation and climate change in Colombia using different econometric techniques. The hypothesis for this research is that innovation plays an important role in supporting consequences of climate change, especially in vulnerable regions. The main contribution of this research is that it applies measured and novel techniques to determine the influence of different variables of innovation on climate change as an input for developing adequate instruments to promote innovation in different Colombian regions.

In general, this study provides an empirical analysis based on a balanced dataset composed of 26 regions or departments and the main city of Colombia from 2016–2018. To assess the relationships between climate change and innovation, this study employs the variables and approach used by global innovation index that was calculated for this country as an input for promoting innovation in whole regions. Therefore, the results of this research have policy implications.

This chapter is structured as follows. The first part provides the introduction. Section two presents the data and method used in this study. In the third section, the results and discussion are explained. Finally, the last section concludes and outlines some suggestions for further research.

Data and Method

In this section, the data and method used in this study are explained, with the aim of determining the relationships between innovation and climate change based on data and variables that are used by the global innovation index and adapted for Colombia to calculate a regional innovation index and analyse how different variables impact innovation performance. This country is selected taking into account improvements in innovation and the development of a new productive economic model that promotes science, technology, and innovation (World Bank 2017).

In Colombia, innovation generates different opportunities for inserting new processes and activities as part of a cumulative flow of economic diversification. Innovation, from incremental developments to drastic originality, can increase the environmental control in this country, which has immensely rich biodiversity and is affected by the stresses created by climate change (OECD 2014). It is important to analyse these relationships to formulate and apply adequate governance and policy to analyse effects and measure climate change.

Data

This analysis uses the Colombian regional innovation index and variables that are based on the global innovation index. The data, variables, and sources of this study are as follows:

- *CO_2 emissions* as tonnes of CO_2 emissions (data on carbon emission factors was obtained from Resolution 181401/2004, where natural gas is 56.1 tCO_2/TJ and electricity is 59.14 tCO_2/TJ); data was obtained from Superintendent of Public Services.
- *Credit for innovation* is measured as percentage of gross domestic product; data was obtained from DANE through the survey on development and technological innovation.
- *Energy efficiency* for every Colombian region; data was obtained from the Superintendent of Public Services and UPME.
- *Sophistication of productive apparatus* is measured by an economic complexity index by region that is calculated based on the formal employment structure of all economic activities that appear in the regional competitiveness index.
- *Industrial specialization* for every Colombian region, which measures regional specialization and shows the degree of similarity of the regional economic structure with that of the country. This measure is generated by DANE through regional accounts.
- *Technology transfer investment* measures investments in technology transfer, the acquisition of licenses, and similar mechanisms that lead to innovation, and is calculated as a percentage of GDP in the data on every Colombian region; data was obtained from the DANE survey on development and technological innovation.
- *Import of high technology goods* in every Colombian region is measured as a percentage of total Colombian imports; data was obtained from DANE's external trade statistics.
- *ICT investment of companies* that introduce new organizational methods; data was obtained from DANE's survey on development and technological innovation.

- *High school education* is measured as the net coverage rate of high school education; data was obtained from statistics published by the Ministry of National Education.
- *Research and development* are measured as investment following Frascati Manual (2015) and the OECD; data was mainly obtained from administrative registers and from the National Planning Department through the regional innovation index.
- *Business expenditures for R&D* as a percentage of sales; data was obtained from the DANE survey on development and technological innovation.
- *Knowledge-intensive employment* is measured as a percentage and is defined as the relationship between the sum of the years of education accrued by the employed population and full employment; data was obtained from the DANE Large Integrated Household Survey.
- *Applications of utility models per* million inhabitants; data was obtained from the national office of the Industry and Commerce Superintendence (SIC).
- *Regional innovation index* for Colombia; data was obtained from the National Planning Department following the methodology of the Global Innovation Index.

Method

This research uses the estimation technique suggested in studies on climate change and interactions with innovation. Different methodologies designed for databases composed of a unique set of cross-section units, in this case, Colombian regions, with an annual periodicity, are employed. With this information, a data panel is constructed for each entity that is studied in each of the reference periods. Three econometric models were adopted with their respective methods based on theory.

Models incorporate corrections recommended by the econometric literature when the time constant causes a residual autocorrelation problem. In the presence of this problem, the methodology of estimation by random effects must be applied, corresponding to a particular case of generalized least squares (GLS). Given the structure of the data, the third model employs a technique that allows us to overcome the typical problems related to the use of longitudinal data. The serial correlation of errors and endogeneity by constant omitted variables in the error can be treated with fixed effects.

The following general structure is considered for estimating random effects:

$$y_{it} = \alpha + \mathbf{x}_{it}\beta + \upsilon_i + \epsilon_{it} \quad t = 1,2\ldots\ldots\ldots,T \tag{1}$$

In this model, $\upsilon_i + \epsilon_{it}$ is the error term that we have little interest in; we want to calculate the estimates of β. υ_i is the unit-specific error term; it differs between units, but for any particular unit, its value is constant.

The estimation structure for climate change is defined as follows:

$$lnCC_{it} = a_{it} + \Phi_1 lnGDPpc_{it} + \Phi_2 lnSPA_{it} + \Phi_3 lnIE_{it} + \Phi_4 lnTTI_{it} + \Phi_4 lnITG_{it} + \upsilon_i + \epsilon_{it} \tag{2}$$

Note that CC_{it} is the Climate Change in period t for Colombian region i, $GDPpc_{it}$ is the aggregate-level production per capita by Colombian region, SPA_{it} is Sophistication of the productive apparatus, IE_{it} is industrial specialization, TTI_{it} is technology transfer investment, and ITG_{it} is imports of high-tech goods.

Similarly, the structure is defined as follows:

$$lnCC_{it} = a_{it} + \varphi_1 lnEE_{it} + \varphi_2 lnSPR_{it} + \varphi_3 lnCIOS_{it} + \varphi_4 lnIICT_{it} + \varphi_4 lnISE_{it} + \varphi_5 lnR\&D_{it} + \upsilon_i + \epsilon_{it} \quad (3)$$

Note that CC_{it} is Climate Change in period t for Colombian region i; EE_{it} is energy efficiency; SPR_{it} is school performance in reading, math and science; $CIOS_{it}$ represents the number of companies certified with ISO quality standards; $IICT_{it}$ is the ICT investments of companies; ISE_{it} is high school education; and $R\&D_{it}$ is research and development.

The following general structure is used for estimating fixed effects for the regression model:

$$y_{it} = \alpha_i + B`x_{it} + \epsilon_{it}$$

A common formulation of the model assumes that differences between units can be captured by differences in the constant term

$$y_i = \boldsymbol{i}\alpha + \mathbf{x}_i\beta + \epsilon_{it}$$

For the innovation model, we estimate the structural form given by the following relationships:

$$lnI_{it} = a_{it} + \Omega_1 lnGDPpc_{it} + \Omega_2 lnCI_{it} + \Omega_3 lnBR\&D_{it} + \Omega_4 lnKIE_{it} + \epsilon_{it} \quad (4)$$

Note that I_{it} is innovation in period t for Colombian region i, $GDPpc_{it}$ is GDP per capita, CI_{it} is credit to innovate, $BR\&D_{it}$, is business expenditures on R&D, and KIE_{it} is knowledge-intensive employment.

Similarly, this structure is defined as follows:

$$lnI_{it} = a_{it} + \vartheta_1 lnGDPpc_{it} + \vartheta_2 lnIICT_{it} + \epsilon_{it} \quad (5)$$

Note that I_{it} is innovation in period t for Colombian region i, $GDPpc_{it}$ is GDP per capita, $IICT_{it}$ is the ICT investments of companies.

Density Analysis

For the analysis of the density estimation, we use the sum of the weighted values calculated with the function of K, according to the following functional form:

$$\hat{f}_k = \frac{1}{qh}\sum_{i=1}^{n} w_i K\left(\frac{x - X_i}{h}\right) \quad (6)$$

The Epanechnikov function is the most efficient way to minimize the mean integrated square error, given the data used for this investigation.

Results and Discussion

Innovation and sustainability have close relationships, and according to their development, they could determine performance, productivity, competitiveness, efficiency, and the generation of new jobs or competences, which require an effective policy and regulation that promote eco-innovation and development with the respective promotion and use of these technologies, products, and goods. In this section, the main results obtained using three econometrics models are analysed. These results are shown in Table 1.1.

Table 1.1. Results of the models used to analyse the relationships among environmental sustainability, climate change, and innovation.

Parameter	**Climate Change**				**Innovation**	
	[1]		[2]		[4]	
	Random Effects	**(Std.Err)**	**Random Effects**	**(Std.Err)**	**Fixed Effects**	**(Std.Err)**
Constant	–15.43 [a]	(6.40)	–0.692	(2.79)	3.498 [c]	(2.56)
GDP per capita	1.457 [a]	(0.39)			–0.024	(0.15)
Credit for innovation					0.035 [a]	(0.12)
Energy efficiency			–0.416 [b]	(0.22)		
Sophistication of the productive apparatus	0.895 [a]	(0.26)				
Industrial specialization	–0.046	(0.26)				
Technology transfer investment	–0.077	(0.05)				
Importations of high-tech goods	–0.033	(0.12)				
School performance, reading, math and science			–0.213	(0.34)		
Companies certified with ISO quality standard			–0.206	(0.17)		
ICT investments of companies			–0.160 [c]	(0.10)		
High School education			2.490 [a]	(0.86)		
Research and development			1.435 [a]	(0.49)		
Business expenditure on R&D					0.024	(0.02)
Knowledge intensive employment					0.060 [b]	(0.03)
F-Model	0.00		0.00		0.00	
F-test for OLS vs. FE	Prob (Reject OLS)		Prob (Reject OLS)		Prob (Reject OLS)	
Breusch-Pagan test (P-value)	0.00		0.00		0.00	
Hausman test (P-value)*	0.1494		0.1452		0.0023	
No. Observations	78		78		78	

Figures in the parentheses are standard errors. [a] Significant at the 1% level, [b] Significant at the 5% level, [c] Significant at the 10% level.

* If Prob > chi2 is < 0.05 reject random effects.

Climate Change Model

Carbon dioxide emissions are the dependent variables in the two models used to analyse the relationship between climate change and innovation. In the first model, climate change has an inverse relationship with industrial specialization, technology transfer investment, and imports of high technologies goods, indicating that decreased CO_2 emissions depend on industrial specialization and the use of new technologies, whereas low formal employment leads to high CO_2 emissions. We take into account that informal employment has not applied environmental protection regulations and protocols.

These results concur with those of Zilberman et al. (2018), who indicated the importance of promoting innovation and new technologies to control and adapt to uncertain climate change impacts that imply improved technology adaptation, the prevention or facilitation of the migration of the population, and increased efficiency. The production sector should analyse the effects of heterogeneous technologies designed for activities, such as pollution control, eco-efficiency, green design, the development of low carbon energy and management systems, to determine the advantages of economic and climate outcomes (Wang et al. 2018).

Moreover, a fully functioning labor market and skilled labor can contribute to a fast, efficient, and fair transition to a low carbon and resource efficient economy, which implies the ability to adapt to new technology because workers are well-performing and have developed strong technology skills (OECD 2012), which must be promoted in school, especially in emerging economies.

The second model indicates that a decrease in CO_2 emissions depends on improved performance in reading, math, and science; the certification of quality standards; energy efficiency, and the ICT investment of companies. Low coverage of high school education and competences in research and development lead to increased CO_2 emissions. These results show that high education levels, certifications in industrial processes based on ISO standards, energy efficiency, ICT investments, and research and development performance are important drivers for controlling and decreasing CO_2 emissions.

These results demonstrate the importance of climate change education, which is an opportunity to enhance interdisciplinary skills that include the complex integration of data, laboratory results, computer modelling, and improved scientific and quantitative analysis based on STEM (McCright et al. 2013). Education on this issue promotes a society with more creative, flexible, adaptable, well-informed, inventive, and sustainable communities (Lehtonen et al. 2019), which is fundamental for climate change adaptation and addressing the challenges of climate change.

Quality standards and infrastructure are essential to check and monitor climate change variables and indicators, which are key elements of generating information for policy-makers and decision makers to formulate adequate governance and policy climate change. This information should include the analysis of the viability, application, and transfer of new technologies; the formulation of national targets and investors' expectations; and the development of quality and safe technologies that can be used to control and adapt to climate change (Ferdinand and Telfser 2017). It is important to promote the application of quality standards in Colombian regions to improve the indicators and measurements and achieve an effective application of technologies and innovation. This is one strategy that can decrease disparities among the territories.

The application of ICT has great potential for reducing greenhouse gas emission through increased energy efficiency, the use of renewable energy, operational process, and others, which could generate large savings and contribute to socioeconomic development (British Telecommunications 2016). It is important to promote infrastructure and the use of ICT, especially in remote areas or regions in Colombia that could promote development and decrease environmental problems.

Controlling climate change is key for the transformation of the global energy system, which highlights the importance of maintaining and conserving the environment, controlling urban air quality, and protecting the world's climate. In addition, energy efficiency offers financial and economic opportunities (IRENA 2019).

The results of these models demonstrate the close relationships among the control or adaptation of climate change and improvements in quality education, new technologies, and innovation, which are fundamental for climate change adaptation.

Innovation Model

In the innovation model, the dependent variable is the regional innovation index based on the global innovation index for every Colombian region. The results indicate a direct relationship between innovation and credit for innovation, knowledge-intensive employment, and the application of utility models, which implies that the government should use different instruments to promote innovation, such as credit, education based on knowledge intensity, and improved innovation processes, including the protection of intellectual property and the register of patents and utility models.

These results demonstrate that science, technology, and innovation play a critical role to prevent, mitigate, and control climate change, which implies that policies for clean technology innovation are designed and applied to decrease the technology gap and increase the role of intellectual property rights in developing countries to achieve a low carbon economy (ICC 2015). Moreover, policies that stimulate demand for innovations that improve efficiency by using alternative energy are important and can lead to sustainability (Gans 2012).

Density Analysis

As noted in the previous sections, this work estimates the different regional distributions in Colombia. This study uses variables related to sustainability, climate change, and innovation to calculate the estimates for the different models. Using the density kernel estimation methodology, it is possible to determine non-parametric density estimators to see how a distribution evolves over time and to analyse the distribution behavior of a variable. Density estimation can show important characteristics of data, such as symmetry and multimodality. By estimating each distribution in isolation and at different points in time, a comparison of the distribution of the analysed variables can be made by observing how they changed between periods (see Figure 1.1).

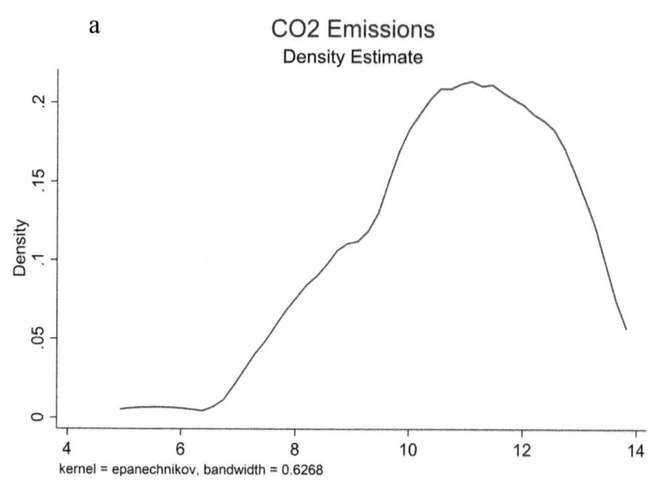

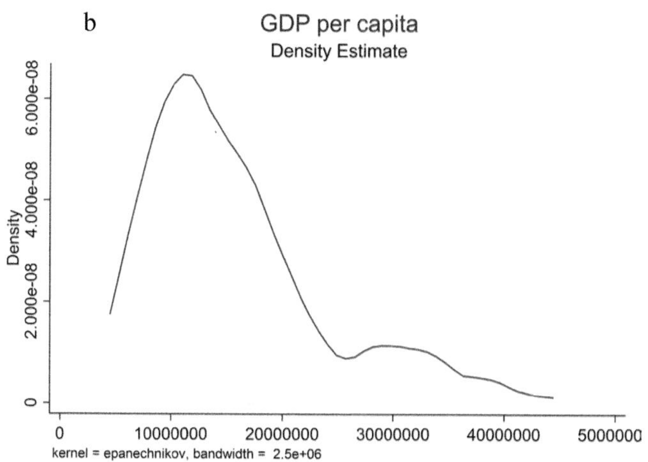

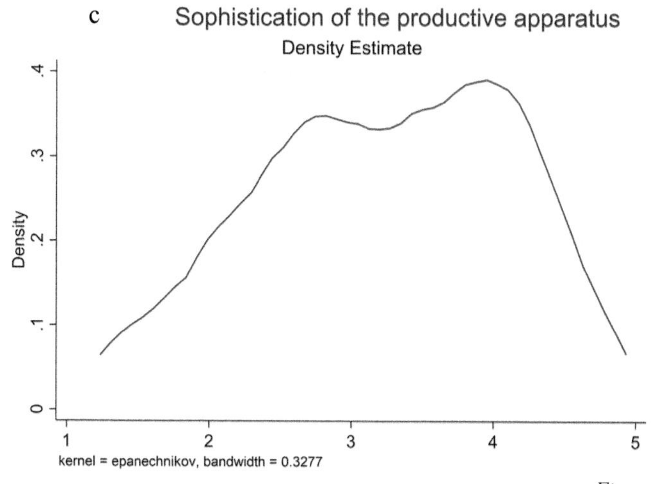

Figure 1.1 Contd. ...

...Figure 1.1 Contd.

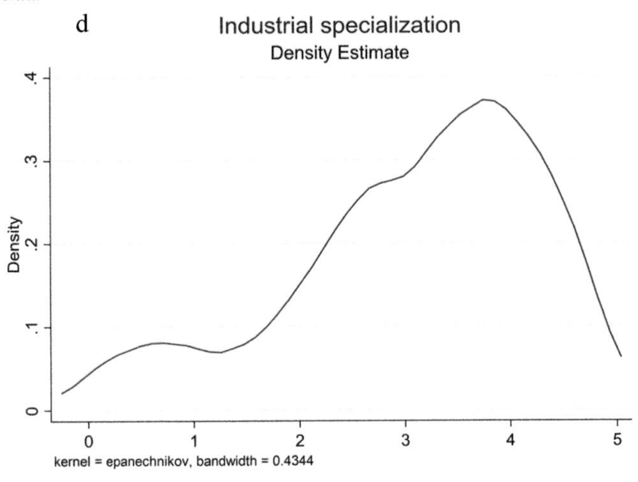

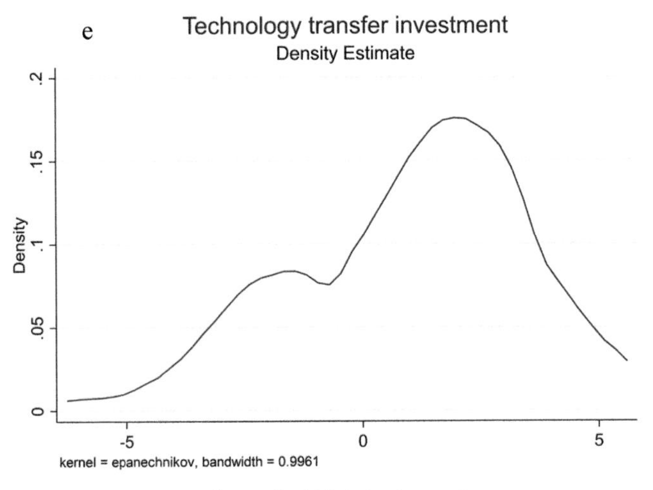

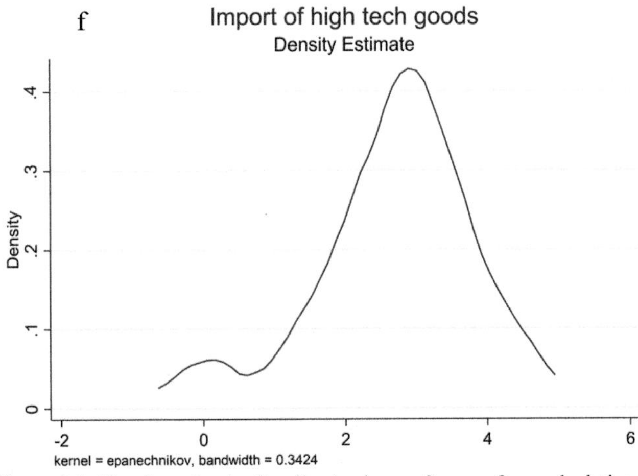

Figure 1.1. Density estimates for climate change. Source: Own calculations.

Figure 1.1 shows a reduction in CO_2 emissions for both low- and high-income regions, achieving outstanding effects on the impacts of climate change for most regions of the country. In the density kernels, the dispersion of the data for the variables of sophistication of the productive apparatus and industrial specialization can be observed, indicating its minimal response to the effects on the reduction of CO_2 emissions. However, the effects of technology transfer investment and the import of high-tech goods show a small dispersion, indicating that they may reduce CO_2 emissions.

Conclusions

Climate change is an important challenge that requires new green technologies, and innovation is a strategy that can help countries migrate to low carbon economies and maintain the standards of living without detrimental consequences for global warming and environmental processes.

The climate change model was worked with two models that were formulated that indicate there is an inverse relationship with industrial specialization, technology transfer investment, and the import of high technologies goods. Moreover, decreasing CO_2 emissions depends on improved performance in reading, math, and science; certification in quality standards; energy efficiency, and the ICT investments of companies.

The density kernels show that the distribution at different periods of time differs, indicating that there is a group of regions that are prone to support environmental sustainability, address climate change, and encourage innovation. The other group of regions should include incentives for these improvements in their policies at the regional level.

The findings of this study are important to formulate and apply adequate innovation and climate change policy that includes clean technologies for mitigation (to reduce the emissions of greenhouse gases) and adaptation (to adjust to the negative effects of climate change or exploit positive ones). In the case of innovation, it is important to engage in technology development (the use of scientific knowledge to obtain solutions) and technology diffusion (new technologies are transmitted from one party to another). More research is necessary to analyse strategies designed to improve innovation and adaptation to climate change, especially in vulnerable areas.

References

Adileri, L., Kotsemir, M. and Vinci, E. (2019). Environmental innovations and productivity: Empirical evidence from Russian regions. Resources Policy 101444 (in press).

British Telecommunications. (2016). The role of ICT in reducing carbon emissions in the EU. https://www.btplc.com/Purposefulbusiness/Ourapproach/Ourpolicies/ICT_Carbon_Reduction_EU.pdf.

Du, K. and Li, J. (2019). Towards a green world: How do green technology innovations affect total factor carbon productivity. Energy Policy 131: 240–250.

EPA. (2018). Reflecting on adaptation to climate change: International best practice review and national mre and indicator development requirements. https://www.epa.ie/pubs/reports/research/climate/research263.html.

Ferdinand, N. and Telfser, K. (2017). Quality Infrastructure and Climate Change in Latin America and the Caribbean. Physikalisch-Technische Bundesanstalt. https://www.ptb.de/cms/fileadmin/internet/

fachabteilungen/abteilung_9/9.3_internationale_zusammenarbeit/docs/PTB_Study_QI_Climate_Change_EN.pdf.

Gans, J. (2012). Innovation and climate change policy. American Economic Journal: Economic Policy 4: 125–145.

ICC. (2015). Supporting innovation to meet climate change challenges. Policy and Business Practices. https://iccwbo.org/content/uploads/sites/3/2015/12/450-1099-ICC-Climate-change-and-Innovation_12-2015_final.pdf.

IPCC. (2014). Climate Change 2014, https://www.ipcc.ch/pdf/assessment-report/ar5/syr/AR5_SYRFINAL_SPM.pdf.

IPCC. (2018). Global Warming of 1.5°C. https://www.ipcc.ch/sr15/.

IRENA. (2019). Climate change and renewable energy. National policies and the role of communities, cities and regions. https://www.irena.org/publications/2019/Jun/Climate-change-and-renewable-energy.

Lehtonen, A., Salonen, A.O. and Cantell, H. (2019). Climate change education: A new approach for a World of Wicked problems. *In*: Cook, J. (ed.). Sustainability, Human Well-Being, and the Future of Education. Palgrave Macmillan, Cham.

McCright, A., O'Shea, B., Sweeder, R., Urquhart, G. and Zeleke, A. (2013). Promoting interdisciplinarity through climate change education. Nature Climate Change Perspective. DOI: 10.1038/NCLIMATE1844.

Morris, D. (2018). Innovation and productivity among heterogeneous firms. Research Policy 47: 1918–1932.

OECD. (2012). The jobs potential of a shift towards a low-carbon economy. Final report for the European commission, DG employment. https://www.oecd.org/els/emp/50503551.pdf.

OECD. (2014). OECD Review of Innovation Policy: Colombia. Overall Assessment and Recommendations. https://www.oecd.org/sti/inno/colombia-innovation-review-assessment-and-recommendations.pdf.

OECD. (2015). Frascati Manual. https://www.oecd.org/sti/inno/Frascati-Manual.htm.

Ranasinghe, A. (2017). Innovation, firm size and the Canada-U.S. productivity gap. Journal of Economic Dynamics & Control 85: 46–58.

Santra, S. (2017). The effect of technological innovation on production-based energy and CO_2 emission productivity: evidence from BRICS countries. African Journal of Science, Technology, Innovation and Development 9(5): 503–512.

Szopik-Depczyńskaa, K., Kędzierska-Szczepaniakb, A., Szczepaniakc, K., Chebad, K., Gajdae, W. and Ioppolof, G. (2018). Innovation in sustainable development: an investigation of the EU context using 2030 agenda indicators. Land Use Policy 79: 251–262.

UNDP. (2004). Reducing disaster risk, a challenge for development. https://www.undp.org/content/undp/en/home/librarypage/crisis-prevention-and-recovery/reducing-disaster-risk--a-challenge-for-development.html.

UNEP. (2018). Innovative solutions for environmental challenges and sustainable consumption and production: concept note on the theme of the fourth session of the united nations environment assembly. http://wedocs.unep.org/handle/20.500.11822/26011.

Wang, D., Li, S. and Sueyoshi, T. (2018). Determinants of climate change mitigation technology portfolio: An empirical study of major U.S. firms. Journal of Cleaner Production 196: 202–215.

World Bank. (2017). Sharing Innovation in Colombia: An Inclusive Approach to Innovation and Competitiveness. https://www.worldbank.org/en/results/2017/10/19/sharing-innovation-in-colombia-an-inclusive-approach-to-innovation-and-competitiveness.

Zilberman, D., Lipper, L., McCarthy, N. and Gordon, B. (2018). Innovation in response to climate change. *In*: Lipper, L., McCarthy, N., Zilberman, D., Asfaw, S. and Branca, G. (eds.). Climate Smart Agriculture. Natural Resource Management and Policy, vol 52. Springer, Cham.

CHAPTER 2

Organic Acid Production by Biotechnology

Lucidio Cristovão Fardelone,[1,*,#] *Taciani dos Santos Bella de Jesus,*[1,#] *Gabriela Chaves da Silveira,*[1,#] *Ynae Padilha David,*[1,#] *Gustavo Paim Valença*[1] and *Paulo José Samenho Moran*[2]

Introduction

Organic acids are molecules that vary in size from small compounds, such as citric and oxalic acids to humic acids, with several carboxylic and phenolic groups and a positively polarized hydrogen atom (McMurry 2008). Organic acids are weak, do not completely dissociate in the presence of water, and have pK_a values ranging from 3 in the carboxylic group, to 9 in the phenolic group (Richter et al. 2007).

Citric, gluconic, itaconic, levulinic, propionic, and acetic acids are part of an important group of basic chemical products because they contain functional groups and exhibit a high transformation capacity. These acids and their derivatives are widely used in the food, chemical, and pharmaceutical industries (Bafana and Pandey 2018, Saha 2017, Krull et al. 2017, Zaman et al. 2017, Cañete-Rodríguez et al. 2016, Pal et al. 2016, Duarte et al. 2015, Ramirez 2015, Weastra 2013, Villanova et al. 2010, Connor 2007).

The global organic acid market was valued at USD 8.27 billion in 2016, and is projected to reach USD 11.4 billion by 2022, at a CAGR of 5.5 percent. The market is growing due to wide-scale use of organic acids in the production of food and beverages, personal care, pharmaceutical and chemical products (Market Research Report 2017).

[1] School of Chemical Engineering, University of Campinas, Av. Albert Einstein, 500, CEP 13083-852, Campinas-SP, Brazil; taciani_bella@hotmail.com, silveira.gabrielac@gmail.com, ynaepd@gmail.com, gustavo@feq.unicamp.br

[2] Institute of Chemistry, University of Campinas, Caixa Postal 6154, CEP 13083-970, Campinas-SP, Brazil; pjsmoran@gmail.com

* Corresponding author: lucfardelone@gmail.com

# Authors contributed equally.

Citric acid production is expected to reach 3 million metric tons and around USD 3.6 billion in 2020 (IMARC 2019). The gluconic acid market generated USD 50 million in 2017 and according to Global Market Insights, USD 80 million is projected by 2024, due to a growing demand for personal care and food products (Globe Newswire 2018). Itaconic acid production is expected to increase from USD 80.8 million in 2016 to USD 102.3 million by 2022, at a CAGR of 4.1%, due to demands in polymeric material applications (Market Research Report 2018a). The levulinic acid market was USD 28.5 million in 2016 and estimated to reach USD 32.5 million by 2021, at a CAGR of 14.0% between 2016 and 2021. This growth can be attributed to pharmaceutical and cosmetic products (Market Research Report 2018b).

Production of other important acids, such as propionic acid, is expected to reach 470 metric tons and USD 1.53 billion in 2020, with animal feed accounting for more than 50% of this total (Grand View Research 2015). Acetic acid is predicted to reach around USD 16.4 billion by 2026 at a CAGR above 6.2% between 2019 and 2026 (Acumen Research Consulting 2019).

Rising demand for organic acids for various industrial purposes has made microbial fermentation an important approach to obtain acids and derivatives from renewable carbon sources (Alonso et al. 2014). The production processes for some organic acids are well known; however, studies are underway on new processes that may be viable for large-scale adaptation (Jang et al. 2012, Sauer et al. 2008). The oldest microbial process for the production of organic acids produces citric acid using the filamentous fungus *Aspergillus niger*, which is capable of producing high volumes at low cost (Berovic and Legisa 2007).

Citric acid ($C_6H_8O_7$), found mainly in citrus fruits, such as orange and lemon, is also produced via the metabolism of fungi, such as *Penicillium citrinum*, *Mucor piriformis*, *Ustilina vulgaris*, *Penicillium luteum*, *Aspergillus clavatus*, and *Aspergillus niger*. Due to its low toxicity, palatability, and safe assimilation by the human body, it is widely used in the food industry as an acidifier, flavorant, antioxidant, preservative, emulsifier, buffer, sequestering and chelating agent, among other applications (Ciriminna et al. 2017, Auta et al. 2014, Costa 2011, Pastore et al. 2011, Connor 2007).

Gluconic acid ($C_6H_{12}O_7$) and its derivatives enhance and preserve the sensory properties of many products, in addition to being considered a weak, non-volatile, and non-toxic acid, capable of forming water-soluble complexes with divalent and trivalent metal ions. It is widely used as an additive in the pharmaceutical, textile, and food industries (Cañete-Rodríguez et al. 2016, Pal et al. 2016).

Gluconic acid, present in foods such as fruits, rice, meat, wine, honey, and vinegar, can be produced naturally through the metabolism of bacteria and fungi, such as *Pseudomonas ovalis*, *Acetobacter methanolicus*, *Zymomonas mobilis*, *Acetobacter diazotrophicus*, *Gluconobacter oxydans*, *Gluconobacter suboxydans*, *Azospirillum brasiliense*, *Aspergillus niger*, *Penicillium funiculosum*, *P. variabile*, and *P. amagasakiense* (Cañete-Rodríguez et al. 2016, Pal et al. 2016, Lopes 2011, Ramachandran et al. 2006).

Itaconic acid ($C_5H_6O_4$) is produced by the metabolic processes of microorganisms such as *Aspergillus terreus*, *Aspergillus itaconicus*, *Ustilago maydis*, *Pseudozyma antarctica*, and *Candida* sp. Its high biodegradation power means it is frequently used in industry to synthesize resins, bioplastics, polyesters, absorbent material, and prepare active compounds for agriculture and medicines (Bafana and Pandey 2018, Saha 2017, Krull et al. 2017, Magalhães Jr et al. 2017, 2016, Ramirez 2015, Pedroso 2014, Takaya 2014, El-Iman and Du 2014, Aşçi and İnci 2012).

Levulinic acid ($C_5H_8O_3$) is a C-5 acid and one of the main chemicals of current interest (Sauer et al. 2008). A weak acid, this compound is obtained mainly via sugarcane biomass hydrolysis and is highly soluble in water, ethanol, diethyl ether, chloroform, and acetone (Mukherjee et al. 2015). Its numerous industrial applications include the synthesis of solvents, pesticides, polymers, polyesters, fuel additives, and pharmaceutical intermediaries (Antonetti et al. 2016).

The production of levulinic acid from lignocellulosic biomass and starch-rich residues is of great industrial interest, since it is environmentally benign and commercially viable (Gaudereto et al. 2017, Kimura et al. 2014).

Propionic ($C_3H_6O_2$), acetic ($C_2H_4O_2$) and succinic ($C_4H_6O_4$) acids are open-chain saturated monocarboxylic acids and among the main products used as intermediates for a large number of products with a wide range of industrial applications, from fine chemicals and pharmaceuticals to the food and cosmetics industries (Caşcaval et al. 2013, Dishisha et al. 2012, Calderón 2012, da Costa et al. 1999).

Propionic acid production uses liquefied petroleum gas by chemical synthesis, employing propanol, propionaldehyde, or ester hydrolysis (Zhu et al. 2012). However, the high cost of these processes, rising liquefied petroleum gas prices, and degree of purity of the desired acid (Wasewar and Pangakar 2006) have prompted the increasing use of fermentative processes (Coêlho 2011).

Thus, all the aforementioned organic acids can be obtained by fermentation, enzymatic, or cascade processes, mixing strains and enzymes. Although current organic acid production processes via these routes are well-established in the industry, there is still a need to develop methods that increase their productivity (Magalhães Jr et al. 2017, 2016, Wasewar and Pangakar 2006).

In acid separation and purification, the perstraction technique can be used in continuous processes to reduce the number of steps and lower production costs. Perstraction involves extraction with organic solvents (namely an alcohol, such as octanol), using a liquid membrane as a barrier, which functions based on the difference in the pH gradient, forming an acid-base complex between an aliphatic amine, such as tri-*n*-octylamine (TOA) and the organic acid. Perstraction has significant advantages over liquid-liquid extraction because it requires less solvent, with minimal loss due to recycling during extraction. Thus, perstraction acts in a three-phase system, where the product is extracted from the aqueous solution (donor phase) through an organic solvent immiscible in water (extractor face) and immobilized in the pores of the membrane, becoming an aqueous solution (acceptor phase) in the lumen of the membrane. The organic phase acts as a barrier between the aqueous phases, that is, the acceptor and donor phases, preventing contact between them (Fardelone et al. 2018, Caşcaval et al. 2013, Oliveira et al. 2008).

In this study, we report the results of organic acid production using a bioreactor coupled to separation membranes, applying the perstraction technique. The use of membranes for separation purposes has grown due to the need for new processes that consume less material and energy, thereby reducing environmental impact (Fardelone et al. 2018, Caşcaval et al. 2013, Oliveira et al. 2008).

General Process

Fermentations were conducted in a fed-batch regime, replacing two-thirds of the culture medium at regular intervals. The fermentations were performed using 3 g of microorganism cells and 0.150 L of culture medium (two-thirds of which was replaced every 24 hours) in a 0.5 L lightly stirred bioreactor.

Acid and glucose concentrations were determined by HPLC with a refractive index detector, coupled to an Aminex HPX-87H column (300 mm × 7.8 mm, Bio-Rad), using a 5 mmol/L H_2SO_4 mobile phase at a flow rate of 0.6 mL/min at 50°C. Acids were quantified from calibration curves using external standards.

Microorganisms

The microorganisms *Aspergillus niger* (a citric and gluconic acid producer), *Aspergillus terreus* (itaconic acid), *Penicillium purpurogenum* and *Penicillium funiculosum* (levulinic acid), and *Propionibacterium acidipropionici* (propionic and acetic acid) were expanded using a culture medium described in the literature, according to each of the strains (Bergey 2000). The culture medium was then optimized for industrial scale, based on commercially available products. All strains are stored in our laboratory, in the form of MCB (Master Cell Bank) and WCB (Working Cell Bank), respectively, and were prepared according to the international standards for good production and laboratory practices (Coecke et al. 2005, Stacey 2004, Health Products and Food Branch Inspectorate 2002).

Preparation of Culture Medium

For the cell expansion (inoculums and pre-inoculums) of microorganisms, the culture medium was prepared with previously established industrial components composed of yeast extract and peptone and 10 g/L of glucose at pH 6.0. A baffled flask was used to obtain high cell density (OD), which promotes better shaker mixing for 18 hours at 120 rpm.

Aseptic conditions were used for all fermentation processes, where 1.5 mL of cells from a Master Cell Bank flask was transferred to a Falcon flask for anaerobic microorganisms, or baffled Erlenmeyer flasks, for better aeration, and kept in a microbiological greenhouse under agitation in the optimal culture conditions for each microorganism (Bergey 2000).

After cell growth, the entire contents of the pre-inoculum flask were centrifuged (3,000 rpm for 5 minutes), the supernatant discarded, and the pellet resuspended in a new culture medium, under aseptic conditions. Cell expansions were conducted in a microbiological greenhouse under agitation, with the optimal culture conditions for each microorganism. After cell growth, the entire contents of the inoculum flask were centrifuged (3,000 rpm for 5 minutes), and the pellet was resuspended in a new culture medium (0.150 L) and transferred to the bioreactor.

Continuous Production System

The continuous production and separation of organic acids by perstraction was initially developed to produce propionic and acetic acids (Fardelone et al. 2018), and subsequently extended for the production of citric, itaconic, gluconic, and levulinic acids, due to its robustness in producing propionic and acetic acids.

The prototype of the production system and continuous separation of acids contains an Infors HT Multifors bioreactor, retention filter, polysulfone hollow fiber membrane, from GE Healthcare (CFP-1-E-3MA), each with a pore size of 0.1 micrometer, area of 110 cm^2 and width of 1 mm, with a total of 13 membranes in the membrane cartridge; two collectors/respirators to retain octanol and remove air from the system, and an extraction container. The system is extremely stable, since extraction with the acceptor phase (NaOH or KOH) is not carried out inside the membrane, according to traditional perstraction, but in a container developed for this purpose, creating two phases, one with octanol and TOA, and the other with NaOH or KOH, as shown in Figure 2.1.

Perstraction uses octanol containing TOA, which is impregnated in the pores and outside the membrane fibers. The donor phase passes through the membrane, allowing the octanol to be recirculated through the perstraction system, without the need to replace parts throughout the process (Fardelone et al. 2018).

The bioreactor system coupled to the perstraction system promotes a closed system process, where the fermentation broth passes through the membranes used with octanol. The acids generated by the fermentative process are extracted through the octanol-TOA system, which forms a complex with the acid produced (Figure 2.2).

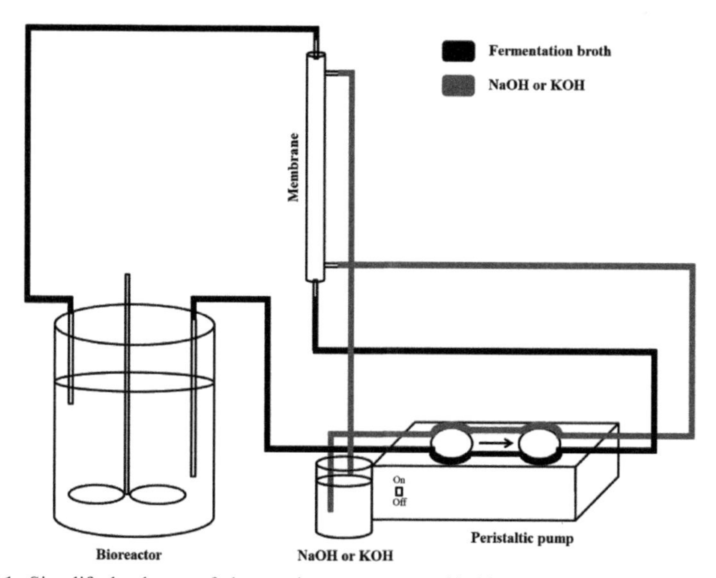

Figure 2.1. Simplified scheme of the continuous system with bioreactor coupled to the separation membrane.

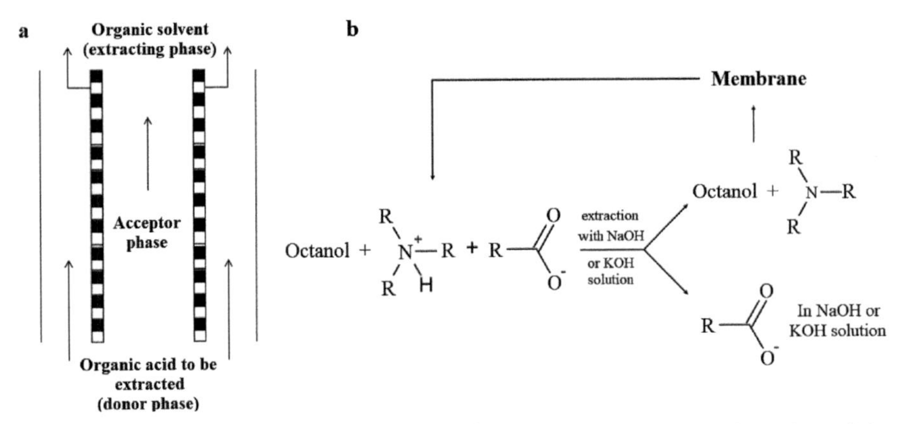

Figure 2.2. (a) Hollow fiber membrane structure for perstraction. (b) Scheme of reactions of the perstraction process.

The extraction container is coupled to the external entrance and exit of the membrane, in order to recycle and after passing through NaOH or KOH, the complex breaks up, generating an accumulation of the desired pre-purified acid. Both the fermentation broth and the octanol solution containing TOA were propelled throughout the system by peristaltic pumps.

During perstraction, the acid is accumulated in a small flask (0.250 L) and the crude product is then purified using an ion exchange column.

Analytical Methodology

For analysis of organic acid concentration, as well as consumption and residual glucose, the analytical methodology described by Coral et al. (2008) was used in high performance liquid chromatography—HPLC, where the detector was the refractive index, with the Aminex HPX-87H column (300 mm × 7.8 mm, Bio-rad) and the mobile phase consisting of a 5 mmol/L solution of H_2SO_4, with flow of 0.6 mL/min and oven temperature of 50°C. However, due to the use of base, such as NaOH or KOH, in the perstraction process, the analytical method was adapted (Fardelone et al. 2018, Bella de Jesus 2016, Coral et al. 2008), since it is necessary to neutralize the samples from the extraction. Thus, the samples will be diluted in a 0.56 M H_2SO_4 solution to neutralize the base, NaOH, or KOH, before the chromatographic runs. The products were quantified from the construction of calibration curves using external standards.

Results and Discussion

Initially, the perstraction experiments involving citric, itaconic, gluconic, levulinic, propionic, and acetic acids have been able to extract acids with weak ionic strength from an aqueous solution with high yields. The molecular structure of the acids under study varied from aliphatic chains, such as acetic and propionic acid to dicarboxylic and tricarboxylic acids (itaconic and citric acids, respectively) and polyhydroxy acid (gluconic acid).

The acid/TOA interaction determines the extraction rate in perstraction. Despite the differences between the pK_as and solubilities of these acids, which may somehow influence the perstraction process, monocarboxylic acids (acetic, propionic, and levulinic) showed higher extraction speed than the itaconic, citric, and gluconic acids (Figure 2.3). The last contain more than one carboxylic group, such as itaconic and citric acids or hydroxyl group such as glucomic acid in their molecular structures.

The interaction mechanism in the formation of an acid-amine complex, where high affinity of the acid to the base provides an additional advantage, is selected over the non-acid components in the mixture. Octanol is a water-immiscible organic solvent with a low solubility of 540 mg/L. Tri-*n*-octilamine exhibits low water solubility and intermediate basicity, allowing the use of a strong base as acceptor phase, such as sodium hydroxide or potassium hydroxide (Wasewar and Yoo 2012, Kaur and Vohra 2010, Oliveira et al. 2008).

Given the good results obtained in perstraction experiments, the continuous acid production process described above was set up on a bench-scale (0.150 L) using the appropriate microorganism for each acid. The yields obtained ranged from 44.7 to 100% (quantitative), generating high purity products (> 99%, after acid purification using perstraction separation and ion exchange resin), Table 2.1.

The productivities obtained were between 0.10 and 0.88 g/L.h, which were dependent on the metabolism of each microbial strain used. Citric and gluconic acid production by the fungus *Aspergillus terreus* and itaconic and levulinic acids by the fungi *Penicillium purpurogenum* and *Penicillium funiculosum*, respectively, are presented in Figure 2.4. A typical propionic and acetic acid production process using the bacterium *Propionibacterium acidipropionici* is shown in Figure 2.5.

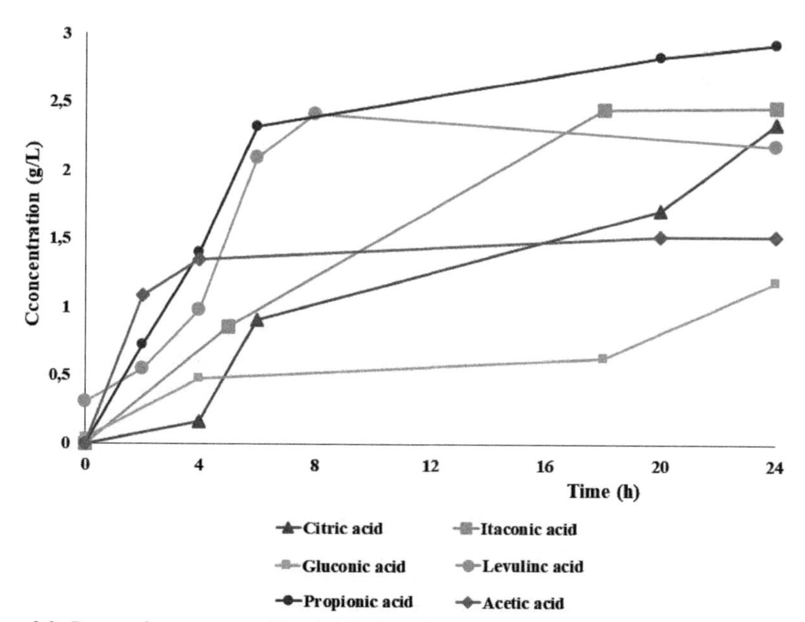

Figure 2.3. Perstraction process profile of citric, itaconic, gluconic, levulinic, propionic, and acetic acids in 24 hours.

Table 2.1. Organic acid production using microorganisms in fed-batch fermentations.

Acids	Productivity (g/L.h)	Yield (%)
Citric[a]	0.43	44.7
Gluconic[a]	0.36	Quantitative
Itaconic[a]	0.88	72.4
Levulinic[a]	0.17	69.1
Propionic[b]	0.23	68.0
Acetic[b]	0.10	73.1

Note: [a] Batch process for 48 hours, due to the metabolism of fungus cells; [b] Batch process for 96 hours, due to the metabolism of bacterial cells.

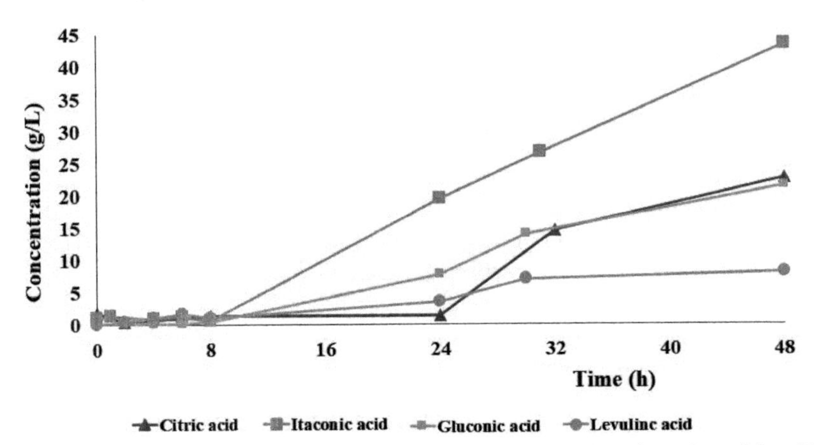

Figure 2.4. Production profile using fermentation of citric, itaconic, gluconic, and succinic acids in 48 hours.

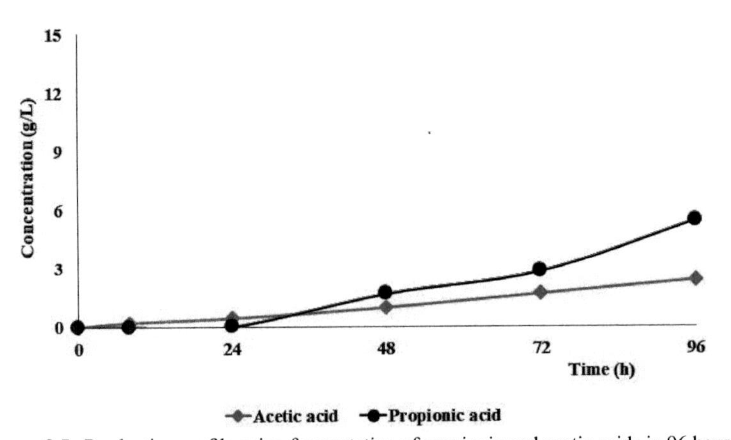

Figure 2.5. Production profile using fermentation of propionic and acetic acids in 96 hours.

The production and separation of organic acids via perstraction is being optimized in order to adapt culture media to larger scales and ensure the process remains robust and viable. It is important to underscore that the polysulfone hollow fiber membrane allows the process to be scaled up. The perstraction process pre-purifies the desired product, after which an ion exchange column is used to obtain the purified acid. In order to scale production, larger capacity bioreactors, equipped with instrumentation for online monitoring of bioproduction are needed, as well as perstraction on a greater scale, using larger membranes.

Conclusions

The bioproduction of the acid under study, along with the perstraction system used to continuously extract the acid from culture medium, was successfully applied to produce citric, gluconic, itaconic, levulinic, propionic, and acetic acids. The continuous organic acid production system described resulted in excellent yields, and may be used to produce other acids.

Acknowledgments

The authors thank the São Paulo State Research Support Foundation (FAPESP, 2016/12074-7), the Coordination for the Improvement of Higher Education Personnel (CAPES, 88882.329694/2019-01 and 147586/2018-6), and the National Council for Scientific and Technological Development (CNPq) for the financial support provided.

References

Acumen Research Consulting. (2019). Acetic acid market size, share and trends analysis report by manufacturing process, applications, by region – global industries size, share, trends and forecast 2019.

Alonso, S., de la Vega, M. and Díaz, M. (2014). Microbial production of specialty organic acids from renewable and waste materials. Crit Rev Biotechnol 35: 497–513.

Antonetti, C., Licursi, D., Fulignati, S., Valentini, G. and GALLETTI, A.M.R. (2016). New frontiers in the catalytic synthesis of levulinic acid: from sugars to raw and waste biomass as starting feedstock. Catalysts 6: 196.

Aşçi, Y.S. and İnci, I. (2012). A novel approach for itaconic acid extraction: mixture of trioctylamine and tridodecylamine in different diluents. J Ind Eng Chem 18: 1705–1709.

Auta, H.S., Abidoye, K.T., Tahir, H., Ibrahim, A.D. and Aransiola, S.A. (2014). Citric acid production by *Aspergillus niger* cultivated on Parkia biglobosa Fruit Pulp. Int Sch Res Not 1–8.

Bafana, R. and Pandey, R.A. (2018). New approaches for itaconic acid production: bottlenecks and possible remedies. Crit Rew Biotechnol 38: 68–82.

Bella de Jesus, T.S. (2016). Separação de ácido propanóico por perstração, obtido por *via* fermentativa, em processo contínuo. M.Sc. Dissertation, Faculdade de Engenharia Química, Universidade Estadual de Campinas, Campinas (SP).

Bergey, D.H. (2000). Bergey's Manual of Determinative Bacteriology. 9th ed. Baltmore, MD: Williams & Wilkins.

Berovic, M. and Legisa, M. (2007). Citric acid production. Biotechnol Annu Rev 13: 303–343.

Calderón, L.A.L. (2012). Estudo de otimização das condições da fermentação para produção de ácido propanoico por *Propionibacterium acidipropionici* utilizando xarope de cana-de-açúcar. M.Sc. Dissertation, Instituto de Biologia da Universidade Estadual de Campinas – Unicamp, Campinas (SP).

Cañete-Rodríguez, A.M., Santos-Dueñas, I.M., Jiménez-Hornero, J.E., Ehrenreich, A., Liebl, W. and García-García, I. (2016). Gluconic acid: Properties, production methods and applications—An excellent opportunity for agro-industrial by-products and waste bio-valorization. Proc Biochem 51: 1891–1903.

Caşcaval, D., Poştaru, M., Galaction, A., Kloetzer, L. and Blaga, A.C. (2013). Fractionation of carboxulic acids mixture obtained by *P. acidipropionic* fermentation using pertraction with tri-n-octylamine and 1-octanol. Ind Eng Chem Res 52: 2685–2692.

Ciriminna, R., Meneguzzo, F., Delisi, R. and Pagliaro, M. (2017). Citric acid: emerging applications of key biotechnology industrial product. Chem Central J 11: article 12.

Coecke, S., Balls, M., Bowe, G., Davis, J., Gstraunthaler, G., Hartung, T., Hay, R., Merten, O.-W., Price, A., Schechtman, L., Stacey, G. and Stokes, W. (2005). Guidance on Good Cell Culture Practice. ATLA 33: 261–287. Available in the Best Practices (GCCP) on the ESACT-UK. The UK Society for Cell Culture Biotechnology website at http://www.esactuk.org.uk.

Coêlho, D.G. (2011). Modelagem e otimização do processo de síntese do ácido propanoico via fermentação do glicerol. M.Sc. Dissertation, Faculdade de Engenharia Química, Universidade Estadual de Campinas – Unicamp, Campinas (SP).

Connor, J.M. (2007). The citric acid industry. In Global Price Fixin, Studies in Industrial Organization book series – SIOR, 26: 113–116.

Coral, J., KARP, S.G., de Souza, V.P., Parada, J.L., Pandey, A. and Soccol, C.R. (2008). Batch fermentation model in propionic acid production by *Propionibacterium acidipropionici* in different carbon sources. Appl Biochem Biotechnol 151: 333–341.

Costa, L.M.A.S. (2011). Caracterização de Isolados de *Aspergillus niger* quanto a produção de ácido cítrico e à expressão de genes da citrato sintase. Ph.D. Thesis, Faculdade de Ciência dos Alimentos, Universidade Federal de Lavras, Lavras-MG.

da Costa, J.P.L.C., Schorm, C., Quesada-Chanto, A., Boddeker, K.W. and Jonas, R. (1999). On-line dialysis of organic acids from a *Propionibacterium freudenreichii* fermentation. Appl Biochem Biotechnol 76: 99–105.

Dishisha, T., Alvares, M.T. and Kaul, R.H. (2012). Batch- and continuous propionic acid production from glycerol using free and immobilized cells of *Propionibacterium acidipropionici*. Biores Technol 118: 553–562.

Duarte, J.C., Valença, G.P., Moran, P.J.S. and Rodrigues, J.A.R. (2015). Microbial production of propionic and succinic acid from sorbitol using *Propionibacterium acidipropionici*. AMB Express 5: 13.

El-Iman, A.A. and Du, C. (2014). Fermentative itaconic production. J Biodivers Bioprospect Dev 1: 1–8.

Fardelone, L.C., Moran, P.J.S., Rodrigues, J.A.R., dos Santos de Jesus, T.B., Valença, G.P. and Nunhez, J.R. (2018). Fermentative process and integrated system for producing organic acids. BR102016031051 and WO2018112577.

Gaudereto, H.S., Cabral, L.G. and Rodrigues, F.A. (2017). Production of levulinic acid from sugarcane bagasse: kinetic study, simulation and economic viability. Engevista 19: 236–255.

Globe Newswire. (2018). Gluconic acid market size by application (industrial [agrochemical & fertilizers, metal surface treatment, textile], beverages, food [confectionary, dairy, flavors, instant food, meat, sauces & dressings], pharmaceutical, personal care, cleaners & detergents), by downstream potential (sodium gluconate, calcium gluconate, potassium gluconate, glucono delta-lactone, industry analysis report, regional outlook, application potential, price trend, competitive market share & forecast, 2018–2024. Global Market Insights, Inc.

Grand View Research. (2015). Global Propionic Acid Market. Grand View Research, Inc.

Health Products and Food Branch Inspectorate. (2002). Annex 2 to the current edition of the good manufacturing practices guidelines schedule drugs (biological drugs) - GUI-0027.

IMARC. (2019). Citric acid Market: global industry trends share size growth opportunity and forecast 2019–2024. IMARC Services Private Limited.

Jang, M.G., Kim, B., Shin, J.H., Choi, Y.J., Choi, S., Song, C.W., Lee, J., Park, H.G. and Lee, S.Y. (2012). Bio-based production of C2-C6 platform chemicals. Biotechnol Bioengin 109: 2437–2459.

Kaur, A. and Vohra, D.K. (2010). Study of bulk liquid membrane as a separation to recover acetic and propionic acids from dilute solutions. Indian J Chem Techonol 17: 133–138.

Kimura, V.T., Poço, J.G.R., Derenzo, S., Matsubara, R.M.S., Gomes, D.Z. and Marcante, A. (2014). Obtenção de ácido levulínico e outros produtos a partir de açúcares usando catalisadores heterogêneos. XX Congresso Brasileiro de Engenharia Química, Florianópolis (SC).

Krull, S., Hevekerl, A., Kuenz, A. and Prüße, U. (2017). Process development of itaconic acid production by a natural wild type strain of *Aspergillus terreus* to reach industrially relevant final titers. Appl Microbiol Biotechnol 101: 4063–4072.

Lopes, S.M. (2011). Estudos preliminares de produção de ácido glucônico a partir de sacarose invertida em biorreator airlift. M.Sc. Dissertation, Curso de Engenharia Química, Universidade Federal de São Carlos, São Carlos.

Magalhães Jr, A.I., de Carvalho, J.C., Ramírez, E.N.M., Medina, J.D.C. and Soccol, C.R. (2016). Separation of itaconic acid from aqueous solution onto ion-exchange resins. J Chem Eng Data 61: 430–437.

Magalhães Jr, A.I., de Carvalho, J.C., Medina, J.D.C. and Soccol, C.R. (2017). Downstream process development in biotechnological itaconic acid manufacturing. Appl Microbiol Biotechnol 101: 1–12.

Market Research Report. (2017). Organic acids market by type (acetic acid, citric acid, formic acid, lactic acid, propionic acid, ascorbic acid, gluconic acid, fumaric acid), application (food & beverages, feed, pharmaceuticals, and industrial), and region – global forecast to 2022. Markets and Markets Research Private Ltd.

Market Research Report. (2018a). Itaconic acid market by derivative (styrene butadiene, methyl methacrylate, polyitaconic acid), application (sbr latex, synthetic latex, chillant dispersant agent, superabsorbent polymer), and region – Global forecast to 2022. Markets and Markets Research Private Ltd.

Market Research Report. (2018b). Levulinic acid Market by application (plasticizers, pharmaceuticals & cosmetics), technology (acid hydrolysis, biofine), and region (North America, Europe, Asia-Pacific and Rest of the World) – Global forecast to 2021. Markets and Markets Research Private Ltd.

McMurry, J. (2008). Organic Chemistry, Seventh Edition, Thonsom Brooks/Cole. United States, Cornell University.

Mukherjee, A., Dumont, M.J. and Raghavan, V. (2015). Review: Sustainable production of hydroxylmethylfurfural and levulinic acid: Challenges and opportunities. Biomass and Energy 72: 142–183.

Oliveira, A.R.M., Magalhães, I.R.S., Santana, F.J.M. and Bonato, P.S. (2008). Microextração em fase líquida (LPME): fundamentos da técnica e aplicações na análise de fármacos em fluidos biológicos. Química Nova 31: 637–644.

Pal, P., Kumar, R. and Banerjee, S. (2016). Manufacture of gluconic acid: A review towards process intensification for green production. Chem Engin Proc: Proc Intensif 104: 160–171.

Pastore, N.S., Hasan, S.M. and Zempulski, D.A. (2011). Produção de ácido cítrico por *Aspergillus niger*: avaliação de diferentes fontes de nitrogênio e de concentração de sacarose. Engevista 13: 149–159.

Pedroso, G.B. (2014). Produção biotecnológica de ácido itacônico a partir da casca de arroz. M.Sc. Dissertation, Curso de Química, Universidade Federal de Santa Maria, Santa Maria.

Ramachandran, S., Fontanille, P., Pandey, A. and Larroche, C. (2006). Gluconic acid: Properties, applications and microbial production. Food Technol Biotechnol 44: 185–195.

Ramirez, E.N.M. (2015). Produção biotecnológica de ácido itacônico por diferentes metodologias de fermentação utilizando o fungo filamentoso *Aspergillus terreus.* M.Sc. Dissertation, Universidade Federal do Paraná, Curitiba (PR).

Richter, D.B., OH, N.-H., Fimmen, R. and Jackson, J. (2007). The Rhizosphere and Soil Formation. The Rhizosphere an Ecological Perspective. Elsevier [s.l.], 179–190.

Saha, B.C. (2017). Emerging biotechnologies for production of itaconic acid and its applications as aplatform chemical. J Ind Microbiol Biotechnol 44: 303–315.

Sauer, M., Porro, D., Mattanovich, D. and Branduarte, P. (2008). Microbial production of organic acids: expanding the markets. Trends Biotechnol 26: 100–108.

Stacey, G. (2004). Fundamental issues for cell-line banks in biotechnology and regulatory affairs. pp. 437–452. *In*: Fuller, B.J., Lane, N. and Benson, E.E. (eds.). Life in the Frozen State. CRC Press LLC, Boca Raton, Florida.

Takaya, R. (2014). Preparação e avaliações comparativas das propriedades físico-químicas entre os hidrogéis de poliacrilato de sódio e de ácido itacônico para potencial aplicação como biomaterial. Ph.D. Thesis, Curso de Ciências Farmacêuticas, Universidade de São Paulo, São Paulo.

Villanova, J.C.O., Oréfice, R.L. and Cunha, A.S. 2010. Aplicações Farmacêuticas de Polímeros. Polímeros: Ciência e Tecnologia 20: 51–64.

Wasewar, K.L. and PANGAKAR, V.G. (2006). Intensification of propionic acid production by reactive extraction: Effects of diluents on equilibrium. Chem Biochem Engin Quarterly 20: 325–331.

Wasewar, K.L. and Yoo, C.K. (2012). Intensifying the Recovery of Carboxylic Acids by Reactive Extraction. 3rd International Conference on Chemistry and Chemical Engineering 38: 249–255.

Weastra, S.R.O. (2013). Determination of market potential for selected platform chemicals: itaconic acid, succinic acid, 2,5-furandicarboxylic acid. BioConSepT, 2013 [Internet]. http://www.bioconsept.eu/wp-content/uploads/BioConSepT_Marketpotential-forselected-platformchemicals_report1.pdf.

Zaman, N.K., Law, J.Y., Chai, P.V., Rohani, R. and Mohammad, A.W. (2017). Recovery of organic acids from fermentations broth using nanofiltration technologies: A review. J Phys Sc 28: 85–109.

Zhu, L., Wei, P., Cai, J., Zhu, X., Wang, Z., Huang, L. and XU, Z. (2012). Improving the productivity of propionic acid with fbb-immobilized cells of an adapted acid-tolerant *Propionibacterium acidipropionici*. Biores Technol 112: 2.

Chapter 3

Urban and Territorial Dimensions of Habitability in Socially Focused Infrastructure Projects

Paula Andrea Valencia Londoño,[a,*] *Diana Valencia Londoño*[b] and *Phoenix Storm Paz*[a]

Introduction

The analytical category of habitability has been primarily used to refer to the architectural analysis of the internal conditions of a residence under the criteria of the right to a dignified life.[1] Sociologists and social psychologists Mercado and González (1991), Mercado et al. (1995) define habitability as the level of satisfaction a person derives from his or her residence, assessing satisfaction based on the building's capacity to respond to the needs and expectations of the inhabitants. Yet, buildings do not exist in isolation; they are part of a cityscape and need to be understood as elements in a larger system. Thinking about buildings as elements in a larger system allows us to extend the category of habitability and assess the outside environment at the urban or territorial level, and to analyze additional variables, such as quality of life, public space, sustainable urban development and growth, social or community participation, healthy living, spirituality, and the search for happiness or personal fulfilment (Morgan and Murgatroyd 1994, Stanley and Dornbusch 1994, Holcombe 1995, Diener 1984). In short, including these categories in the evaluation

[a] University of Medellin, Carrera 87 N° 30 – 65, Medellin, Colombia; phoenix.storm.paz@gmail.com

[b] University of San Buenaventura, Carrera 56C N° 51-110 Centro, Medellin, Colombia; diana.valencia@usbmed.edu.co

* Corresponding author: pvalencia@udem.edu.co

[1] Article 51 of the Colombian Constitution of 1991 guarantees citizens the right to a residence that provides a minimal standard of dignified living conditions. The article reads: "All Colombians have the right to a dignified residence. The State will determine the conditions of quality of life needed to affect this right, promote plans that integrate living conditions with social interests, provide adequate financing in the long term, and put in place social programing in relation to the maintenance of these conditions."

of habitability relates the concept once more to the human subject, offering a look at how a subject responds to his or her environment, where the constructed urban environment is understood in relation to human necessities (Moreno Olmos 2008).

In this chapter, we ask how the theoretical concepts of internal and external habitability can be applied to analyze the impacts of nine urban infrastructure projects built as a part of the public policy of "Social Urbanism" in Comuna 8[2] of Medellin, Colombia. Our objective is to demonstrate the insufficiency of using internal habitability alone as a measurement of quality of life. To do so, we divide the chapter into four sections.

The first is a theoretical discussion of internal and external habitability. In this discussion, we break away from the existing school of thought in two main ways: first, we expand the scope of habitability to understand the entire city and territory as a single integrated system that must be considered as a whole and second, we expand the purpose of the habitability discussion from speaking about the individual to speaking about society, or groups within a society. These two breaks endow the concept with "the capacity to assess social and environmental issues" (Arcas-Abella et al. 2011). After discussing the definition of the concept, we address the techniques used for measuring habitability, suggesting that the analysis of habitability should include both qualitative and quantitative evaluations. Quantitative measurements give us a comparative value to use in rating of different urban designs. However, evaluating the physical elements of urban space is insufficient; the assessment of habitability must also include the qualitative study of the social networks that contribute to a sense of belonging to the place a person inhabits.

In the second section we approach the case study, providing historical, economic, and social context for Medellin, Colombia and the political paradigm of "Social Urbanism", explaining how it was carried out in Comuna 8 where more than 500 urban infrastructure projects have been built in the last decade. We offer brief descriptions of the nine urban intervention projects chosen for this study, which include: eco-parks Villa Turbay, Las Tinajas and 13 de Noviembre; the UVAs Santa Elena and Sol de Oriente; the library-park León de Greiff; the recreational park Las Estancias; and the urban hiking and biking trails El Camino de la Vida and la Ruta de Campeones.[3]

In the third section, we provide the methodology of our case study, explaining the instruments applied in the collection of primary and secondary information. Primary material was collected in the research project "A Comparative Analysis of the Identification of the Social, Environmental and Territorial Impacts of Urban Projects",[4] in which a matrix of indicators was developed to quantitatively measure

[2] In Colombia, municipalities have urban and rural zones. Urban zones are divided into administrative units called "comunas". Medellin has 16 comunas. Rural zones are called "corregimientos", of which Medellin has 5.

[3] These trails run the perimeter of the city; the first is primarily a hiking trail, and the second a biking trail. They mark the boundaries between the urban and rural zones in the municipality of Medellin.

[4] This project was developed and funded in partnership between three universities located in Medellin, Colombia: la Universidad de Medellin, el Colegio Mayor de Antioquia, and la Universidad de San Buenaventura.

habitability in relation to the impacts of urban renovation and development projects on urban quality of life.

Finally, in the fourth section, we analyze the results of the primary research and put these results in dialogue with the secondary information collected from official sources. We delve into four specific aspects of the intervention projects we feel best represent the integrated facets of external habitability at the territorial scale: qualitative housing deficiencies, public space, green space, and territorial sustainability. For each, we put biophysical information from primary and secondary resources in dialogue with sociocultural indicators from our primary research. In doing so, we problematize the common definition of habitability as an architectural analysis of the physical conditions of a residence, pointing out the gaps in the official studies of conditions in Comuna 8. The correspondences and discrepancies that arise when comparing these different sources of data at different scales demonstrate the importance of an expanded notion of habitability. We close the chapter with a series of conclusions based on the analysis discussed in the findings and discuss some of the limitations of the project.

External Habitability as a Theoretical Framework

The concept of habitability was first used in México by Mercado and González (1991), to describe a subject's psychological response to their living situation and the conditions of their surrounding environs (Mercado and González 1991). They suggest that habitability can be explained through variables such as the enjoyment of, privacy in, and psycho-emotional response to a place (Mercado et al. 1995). Mercado et al. (1995) defined the concept as the level of pleasure that the inhabitants of a particular household receive in the fulfilment of their necessities or expectations, building off a matrix evaluation technique offered by Kelly (1955) to offer a means of measurement.

Yet, despite its origins in sociology and social psychology, habitability is used principally to establish legal norms and regulations around building and construction practices, particularly in countries such as Spain (Casals-Tres et al. 2011) and Colombia. In these countries, habitability analyses are used to ensure that residential buildings comply with the constitutional guarantees of the right to a dignified life, but are limited to architectural analysis. This type of architectural habitability in the interior of a residence has been labeled by scholars as "internal habitability".

In contrast with internal habitability, the works of Mercado et al. (1995), as well as Landázuri and Mercado (2004) consider habitability from outside of the residence, introducing a concept they call "external habitability". This last perspective refers to the systemic level in which the immediate urban environment is considered when evaluating whether a residence is habitable or not, building off the ecological model proposed by Bronfenbrenner (1971). External habitability seeks to understand the interconnectivity between residence, neighborhood (Moreno Olmos 2008), urban infrastructures and installations, and the use of these spaces by city residents (Páramo et al. 2018). The merger of internal and external habitability allows each scale of territory to be evaluated in its capacity to satisfy human necessities (Moreno Olmos 2008).

Castro (1999), Landázuri and Mercado (2004), and Moreno Olmos (2008) define habitability as the satisfaction obtained in a specific space, or the capacity of built spaces to satisfy the objective and subjective needs of the individuals and groups that occupy said spaces. This implicates an essential redefinition "of the concept of habitability that allows it to face social and environmental issues" (Arcas-Abella et al. 2011), with the end goal of "an integrated understanding of the system of structures and actions that allows the satisfaction of social needs, regulates the taking of materials from the environment, and stimulates strategies of transformation" (Arcas-Abella et al. 2011). The expansion of the concept of habitability shows that, as Villagrán Garciá (1963) affirms, the habitable necessarily implicates the relationship between human beings and space.

Understanding habitability as the relationship between subjects and their environments allows different authors to explore the concept and endow it with different focuses. Moreno Olmos (2008), for example, focuses on quality of life. In contrast, Levy and Anderson (1980) focus on socioeconomic class, Dasgupta (1988) on well-being, and Velásquez (1988) on urban life practice. Morgan and Murgatroyd (1994), Stanley and Dornbusch (1994), Holcombe (1995), and Diener (1984) suggest that habitability includes everything from social and community participation to the practice of a healthy life, to spirituality, and the satisfaction of personal or individual needs for the realization of happiness. In contrast, Drucker (1983) sees habitability as a means of measuring demographic expansion to productivity and the processes and conditions in the workplace, while Clarke (1977) relates the concept to economic growth. Other authors understand habitability in relation to the proposal by Max-Neef et al. (1986) of human necessity, defining the concept as the relation between humans and their environments, and the adequacy thereof.

Moreno Olmos (2008) defines four ways of understanding habitability. First, habitability is an intangible, qualitative conditional that relates to the human soul and well-being. Secondly, habitability is a quantitative measurement of quality of life, which can be used to drive action in urban planning and transformation. This action is quantifiable and controllable and includes improving spatial conditions under the guidance of standards determined by specialists. Thirdly, habitability can be understood as a tool for evaluating living conditions and residential comfort. Lastly, Moreno Olmos (2008) suggests that habitability is a perceptive act, an interpretation of the ways in which people occupy and inhabit space, and of their physical expressions of the psychophysical world in which they reside.

The second definition of habitability as quantitative measurement related to "Quality of Life" (QoL) analyses is the most similar to the concepts presented in the English-language literature. According to Jimmy et al. (2019), QoL refers to the level of satisfaction that residents feel in relation to their neighborhood, specifically based on their experiences within the place they inhabit (Jimmy et al. 2019, Senlier et al. 2008). The English-language QoL literature evaluates how the physical structure of the space responds to the psychosocial needs of the city inhabitants on a variety of levels (Jimmy et al. 2019) and how people relate to their everyday urban environments (Jimmy et al. 2019, Pacione 2003). Relating habitability to QoL calls for an evaluation of the subjective components—psychological and social—of city life (Jimmy et al. 2019, Santos et al. 2007, Senlier et al. 2008, Shumi et al. 2015, Sirgy et al. 2008) as well as physical components, including both the internal that relates to specific physical structures and the external that relates to the integration of public spaces and public services (Jimmy et al. 2019, Berhe et al. 2014, Martinez et al. 2016, Tesfazghi et al. 2010). The integration of the two levels of analysis corresponds to the definition of habitability proposed by Spanish-speaking authors who refer to "wellbeing, deprivation, adaptation and dissonance" when speaking of QoL (Jimmy et al. 2019, Berhe et al. 2014, Craglia et al. 2004, Tesfazghi et al. 2010).

Because the discussion of habitability includes discussion of human beings' psychological and emotional responses to their environments, it therefore refers not only to the physical stability of a building and its relationship to its physical environment (both internal and external), but also to its cultural environment. Korean architects Jeong et al. (2019) argue that architecture "has the duty to preserve the past and provide the possibility of building the present on the strength of culture and tradition" (Jeong et al. 2019). Their comments emphasize the importance of the concept of urban QoL, which can be defined as "the optimal conditions that combine to determine the biological and psychological aspects that constitute the enjoyment of a space that human beings inhabit and in which act" (Pérez Maldonado 1999). In the case of the city, these conditions have to do with the level of satisfaction with public services and the "perceptions of habitable spaces as clean, safe, healthy, and visually stimulating" (Pérez Maldonado 1999). Nonetheless, according to Páramo et al. (2018), the concept of urban QoL should be expanded to associate conditions of comfort or enjoyment of a space with the "ecological, biological, economic, productive, sociocultural, typological, technological and aesthetic in all their spatial and psychosocial dimensions." All that, for the most part, is very similar in meaning to definition of habitability first proposed by Mercado et al. (1995) with respect to the psychological indicators of perception.

From an expanded perspective that understands habitability from the proposal by Max-Neef et al. (1986) of human necessity, the relationship between ecological, biological, economic, productive, sociocultural, typological, technological, and aesthetic environments permits scholars to think about an urban QoL that moves beyond the residence to include a more expansive gamut of satisfiers and necessities, referring not only to the domestic household level, but also to the entire city. The purpose of using an expanded urban QoL framework is that it allows us to question-through an evaluation of enjoyment and appropriation—whether a space responds to the functions required from it (Jeong et al. 2019), and whether a place can fulfil different social roles and provide the various types of well-being that inhabitants desire in order to transform the city.

Arcas-Abella et al. (2011) specify that "a model of habitability inconsistent with the large variety of demands that it should respond to, can only with great difficulty respond with an appropriate, proportional answer. Inadequacies in habitable conditions can lead to inappropriate consumption, and therefore inefficient use of resources, offering solutions that do not resolve the problems." Arcas-Abella et al. (2011) and Casals-Tres et al. (2011) therefore argue that it is necessary to establish habitational conditions coherent with the right to a dignified life and household, as well as the challenge of sustainability, due to "the current definition of habitability as interdependent with the resources used and the waste products produced in the construction and maintenance of a place in time" (Arcas-Abella et al. 2011, Casals-Tres et al. 2011).

The concept of sustainable urban construction and development can be one way to find the intersection between habitability and sustainability, as it refers to the extrapolation of the concept of sustainable development for the satisfaction of human necessity, where the careful use of resources can result in more habitable living conditions. According to both Arcas-Abella et al. (2011) and Casals-Tres et al. (2011), habitability should be redefined in relation to environmental impact. Habitability should transcend the domestic sphere to speak to the urban scale when responding to the social demand for particular living conditions, taking into consideration that the sustainability of domestic activities is dependent upon questions at the urban and territorial levels.

Further developing the argument, Casals-Tres et al. (2011) affirm that the recognition of the urban scale of habitability allows for the following: first, the consideration of a social dimension in habitability; second, the generation of an integrated vision of habitability coherent with sustainability; and third, the provision of the tools needed for the transformation of the built environment.

In that sense, urban habitability makes reference to what Bentley et al. (1985) call, among other aspects, "the study of the qualities that are developed in the urban environment outside of the architectural spaces" (Moreno Olmos 2008). Alcalá Pallini (2007) "conceptualizes habitability in urban spaces as the habitational condition where the residence is physically integrated with the city, with good access to services and recreational structures, surrounded by a high-quality public space. The lack of accessibility renders an area vulnerable or marginalized even when the residences are in good condition". This definition of habitability obligates us to analyze urban policies in spaces between neighborhoods; in the roadways and infrastructures; public transit and green spaces; new centralities and common places of reference to determine its direct impact on the improvement of living conditions (Moreno Olmos 2008); and finally, on the quality of urban life.

A clear example of the merger of the two modes of thinking about habitability as a response to the challenges of sustainability at the urban level are the studies carried out by Alcalá Pallini (2007), Marquina and Pasquali (2006), Páramo and Burbano (2018), which analyze public space in Latin American cities as a condition of habitability. Public spaces are places where social practices are developed and performed (Páramo et al. 2018). Moreno Olmos (2008) argues that the city itself is a public space where collected living is organized according to the need for social networks. These networks, in turn, improve the level of habitability and generate the feelings of belonging associated with inhabiting a space physically and socially. Public space impacts urban quality of life in relation to the satisfaction of human necessities; it, therefore, should be habitable.

Páramo et al. (2018) argue that habitability, understood as access to public space and public life, is "one of the fundamental rights of the citizens in relation to their city". They show the importance of public space in the habitability of a neighborhood through a study conducted in 11 cities across seven Latin American countries. Like Marquina and Pasquali (2006), the authors affirm that clean walkable streets and the presence of nature improve the levels of satisfaction with public space, and are, in fact, the most important factors in deciding an inhabitant's response. There is an importance given to "aspects like physio-spatial structure, design and quality of the environment" (Páramo et al. 2018).

The emphasis on the importance of public space to public life offered by Páramo et al. goes in contradiction to the typical disinterest observed in the academic analysis of the topic, which is perhaps due to the fact that "urban public space is not considered as habitable space, but rather a place of 'transit' where habitability does not come into play" (Valladares et al. 2015).

To conclude, there is a lack of a uniform global conception of habitability, which calls for a redefinition that "should start from the consideration of habitability as the response to the social demand for accessibility and the availability of precise utilities to satisfy socially defined living conditions. It should also consider the natural resources needed in obtaining said living conditions and regulate the waste produced in maintaining them" (Arcas-Abella et al. 2011). This redefinition provides the criteria for the integrated evaluation of urban QoL under the umbrella of habitability, a concept which recognizes the processes of integration and fragmentation of communities caused by the physical structures that give rise to social structures.

The previous discussion shows how, by engaging with the English-language QoL literature as well as Spanish-language habitability literature and arguing for public space as inhabitable, we break away from the existent Spanish-language habitability literature in two main ways: the first in relation to the scale of habitability analysis and the second in respect to aim, from a merely physical analysis to one that engages both physical and social dimensions of quality of life.

Historical, Economic, and Sociopolitical Context

Medellin: A History of Migration and a Multiplicity of Violences

Medellin's story is one of migration. Until the end of the 19th century, Medellin remained a small mining town nestled in the Valley of Aburrá[5] in the western mountains of Colombia, largely inaccessible. However, with the construction of the railroad in the mid-19th century, the city's fortunes changed and Medellin became the center of Colombian industrial.

The end of the nineteenth century and the first decades of the twentieth century bore witness to mass migration from rural areas, as workers flocked to the city seeking employment in the burgeoning textile industry. Consequently, the demand for housing skyrocketed. Housing initiatives were launched through private inversion, and various neighborhoods were built to house the new working classes (Valencia 2017).

[5] The Valley of Aburrá is home to Medellin and nine other municipalities that make up "The Metropolitan Area of the Valley of Aburrá" (el Área Metropolitana del Valle de Aburrá). With an extension of 1,152 km^2, the valley is better described as a canyon. Only 340 km of valley floor are urbanized (Cámara de Comercio de Medellin para Antioquia [n.d.]), home to some 3,726,219 inhabitants according to the 2018 census (DANE 2019).

A second wave of mass migration to Medellin occurred in the 1950s, with vastly different characteristics from the first. Rather than the economic migration of the late 1800s, the migrants of the 1950s were refugees forcibly displaced by La Violencia, a period of extreme socio-political violence between liberals and conservatives.

In this period, Medellin's population doubled, with an annual growth rate of more than 6% (Echeverry and Orsini 2010). Informal settlements and invasions arose in the peripheries to the north of the city, and along the western and eastern hills of the valley in lands that are geographically complicated, fragile, and difficult to access. The municipal administration of the period proved insufficient to respond to the huge population growth, and private companies were uninterested in building neighborhoods for people who were not their employees, resulting in a dearth of infrastructure and public services in areas where refugees settled and built their homes themselves (Valencia 2017). Today, approximately 50% of Medellin's population lives in these peripheral areas (Echeverry and Orsini 2010).

In sharp contrast with the hillsides, the center and south of the Valley of Aburrá are populated by the middle and upper classes with formal, planned neighborhoods. As such, Medellin is divided in "two opposing 'cities' dramatically segregated by location and geographic conditions" (Echeverry and Orsini 2010).

The 1960s saw the rise of politically left guerilla groups, a legacy of the socio-political violence of the previous two decades. While initially located in rural areas of the country, these groups expanded to urban hubs, including Medellin, throughout the 1970s. The National Liberation Army (Ejército de Liberación Nacional or ELN), for example, founded urban militias as a means of consolidating territory and securing provisions. However, cities were not merely supply points, but rather, also places for expanding the group's support base and making alliances with other social movements including syndicates. With their insertion into urban areas, guerilla militias began offering neighborhood security, gaining both legitimacy and territorial and social control, and started administering justice in competition with local and national institutions (Medina Gallego 2001, Valencia 2018).

In response to the presence of guerillas, paramilitary groups arose in the 1980s, often linked to organized crime associated with drug trafficking. For example, La Terraza was a paramilitary assassins guild operating between 1984 and 1998 that monopolized armed action on behalf of the Medellin Cartel (el Cartel de Medellin), Pablo Escobar's narcotrafficking empire. The presence of multiple actors in the city gave rise to a "combination of different forms of organized violence: the simultaneous action of narcotraffickers, militias, and organized as well as unorganized crime–gangs" (Defensoría del Pueblo 2004).

The heyday of this brand of multi-actor violence spanned from 1990 and 1994, when the city was occupied by FARC and ELN urban militias, as well as paramilitary groups affiliated with the United Colombian Auto-defense Force (Autodefensas Unidas de Colombia or AUC), due to the strong international persecution of the Medellin Cartel. Between 1995 and 1997, the government established the Community Watch and Services Cooperative (Cooperativa de Vigilancia y Servicios Comunitarios or Coosercom), which founded additional private security paramilitary groups to control crime rates (Defensoría del Pueblo 2004).

The fight between the guerillas and the new paramilitary forces sponsored by the government peaked between 1999 and 2003, ending with the final takeover of the city by paramilitary groups. Former urban militia members and civilians living in the disputed territories found themselves the victims of violent attacks. Operations Mariscal and Orion carried out in Comuna 13 in the year 2000 are examples of the "new manifestation of the collaboration between the central government and criminal bands. The apparent objective was to 'civilize' criminal activity or 'zero tolerance' for organized crime" (Palacios and Serrano 2010).

Medellin, the second largest population receptor for people displaced by the Colombian Armed Conflict in the late 1990s and early 2000s, transformed into a stewpot for the confluence of various armed actors (Valencia 2018). Due to the resurgence of the armed conflict between guerillas and paramilitaries in rural areas, inner-city violence returned with a new twist: forced displacement within the city itself. Present in Medellin since 2008, inner-city displacement grew 300% annually until 2011, when growth rate dropped to 60%, according to the Personería de Medellin (Valencia 2018).

The processes of spontaneous and unorganized growth, products of the continued arrival and forced relocation of populations already displaced by rural violence, result in a persistent inequality that continues to characterize Medellin (Tilly 2003). According to the "How are we, Medellin?" (Medellin Como Vamos) quality of life report, inequality was reduced by a mere 0.03 points from 2010 to 2017, dropping from 0.55 to 0.52 over seven years. These statistics demonstrate a startling stagnancy in the face of international standards proposed by the UN Objectives of Sustainable Development for 2030, and contradict the positive changes reported by the Metropolitan Area in reduction of inequality. The difference in interpretation is largely due to what information an organization chooses to analyze, the institutional and structural inequalities, or the changes in income distribution (Medellin Como Vamos 2019).

"Social Urbanism": A Palliative Measure Tested in Comuna 8

In an attempt to reduce the inequality gap segregating the city, the municipal administrations of Sergio Fajardo, Alonso Salazar, and Anibal Gaviria, spanning from 2004–2016, designed and implemented a public policy called "Social Urbanism". The policy adopted a strategy of urban renovation and large-scale inversion in urban infrastructures in impoverished sectors of city to add social and aesthetic value to the territories in which they were built, and thereby diminish social differences.

Through this policy, Medellin transformed its international reputation from a city of violence to an innovative city at the head of Latin American development. The city has won multiple awards attesting to this fact. In 2013, it won the Wall Street Journal and Citi Group's "City of the Year Award". For innovation and daring sustainable development, Medellin won the 2016 Lee Kuan Yew World City Prize. In 2017, it won the national prize for digital innovation in mobility, and in 2018, the Autodest Prize for Excellence for development designs proposed by the NGO United for Water and the Integrated Neighborhood Improvement (Unidos por el Agua y Mejoramiento Integral de Barrios). Finally, in 2019, Medellin won the Innovative

City of the Year Prize from Nearshore Americas for impact in science, technology, and innovation programming.

Comuna 8 of Medellin, located in the central-eastern sector of the city, was the principal laboratory for "Social Urbanism" policies, particularly during the administration of Anibal Gaviria (2012–2016). The comuna was populated in the mid-20th century, when economic migration gave way to forced displacement related to La Violencia. While the first settlements were built by Coltejer Textiles for company employees in the foothills behind the plant in the 1940s, the higher slopes were populated by rural migrants displaced by the sociopolitical violence of La Violencia throughout the 1950s. The land was parceled into lots which were urbanized through legal and illegal construction, resulting in the almost over-night appearance of neighborhoods.

The current population is approximately 135,000 inhabitants, the majority of whom are between 15 and 45 years old. This comuna is characterized by informal settlements (there are five legally recognized neighborhoods and 16 irregular settlements) and marked by poverty (about 34,000 of the 46,000 households in the comuna belong to socioeconomic strata 1 and 2).[6] Moreover, the comuna occupies

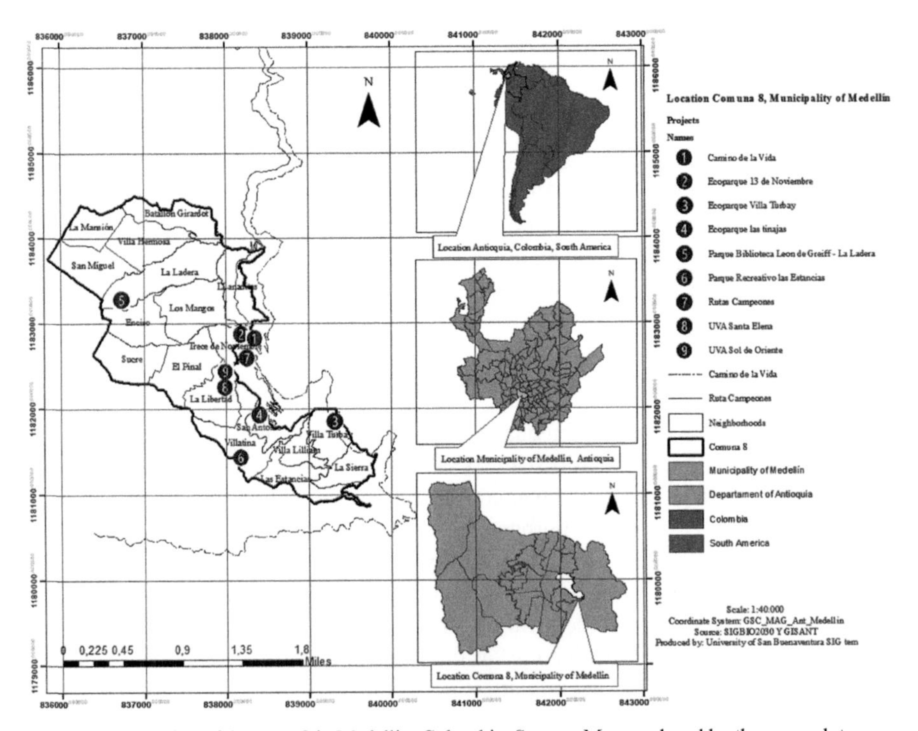

Map 3.1. Location of Comuna 8 in Medellin, Colombia. Source: Map produced by the research team.

[6] Colombia has a stratified socioeconomic model that uses a scale between 1 and 6 to classify residences according to their socioeconomic characteristics. The initial purpose of this model was to charge a lesser rate for public services to households with greater economic needs. Now, it is also used to offer subsidies and other government help.

the 4th place in the six worst-ranked sectors on the City Quality of Life Index (Indice de Calidad de Vida de la Ciudad or ICV) in the period of 2010 to 2018. In eight years, the quality of life index increased by only 3.2 points, 36.5 to 39.7 (out of a total 100). The difference between Comuna 8 and the best scoring sector in the city, El Poblado, which has a score of 76.6 points, is nearly 30 points (Medellin Como Vamos 2019).

Urban Interventions Chosen for Analysis

In contrast with the limited advance in quality of life, the administration of Anibal Gaviria (2012–2016) built nearly 500 different infrastructure projects as part of the "Social Urbanism" policy, which can be categorized into six main types of interventions: public space; social facilities including playgrounds and outdoor gyms; consolidation and formalization of settlements; expansion and improvement of public service infrastructures; improvement and expansion of the roadways; and environmental recuperation and renewal. Of all the projects, nine stand out and form the basis of the analysis in this paper.

Ecoparque Villa Turbay

Villa Turbay is one of two eco-parks built for the Medellin Circus Garden (el Jardín Circunvalar de Medellin). Although located in Villa Hermosa, a neighborhood in Comuna 8, Villa Turbay connects to the La Sierra and Media Luna neighborhoods as well. The eco-park consists of ecological corridors and hiking paths, children's playgrounds and outdoor gyms (Cinturón Verde [n.d.]).

Ecoparque Las Tinajas

Ecoparque Las Tinajas was built to reduce environmental risk in the La Sierra neighborhood of Comuna 8 and offer the community a sports facility, remodeling of the La Granja soccer field. Some 8,000 square meters of the park's total 23,500 square meter area were transformed into public space, which includes a synthetic field, an INDER office, an outdoor urban gym, a children's playground, an urban vegetable plot garden, a plaza, and parking lots (EDU 2014).

Ecoparque 13 De Noviembre

Ecoparque 13 de Noviembre was built to recoup the Fosforite sportsground, which was in total ruins and transform it into a tourist attraction. Improvements consisted of enclosing the field and building grandstand seating. Six plazoletas consisting of vista points, outdoor gyms, children's playgrounds, and other tourist attractions were also built. Various archeological sites were uncovered in the construction process. As in Eco-park Las Tinajas, one of the goals for Eco-park 13 de Noviembre was natural risk mitigation. To that end, more than 160 meters of the stream La Arenera were channeled and bridges were built at two points to facilitate pedestrian access (EDU [n.d.]a).

UVA[7] Santa Elena

UVA Santa Elena in located in a renovated aqueduct and telecommunications infrastructure. Complementary structures that maintain the cylindrical form of the existing water tank were added. Neighborhood inhabitants use the UVA's community classrooms, computer rooms as a public service access point (EPM 2013).

UVA Sol De Oriente

UVA Sol de Oriente was developed by the Urban Development Company (Empresa de Desarrollo Urbano or EDU), Medellin Public Works (Empresas Públicas de Medellin or EPM), and the Medellin Sports and Recreation Institute (Instituto de Deporte y Recreación de Medellin or INDER). The project included the rehabilitation and remodeling of the Sol de Oriente sports field, formerly a sand soccer field, converting it into a synthetic field located on the 3rd floor of a multipurpose facility built into the hillside. The other floors of the UVA are comprised of new community spaces, including multipurpose rooms, events halls, computer rooms, water playground with fountains and ponds. In the associated grounds, there are hiking trails, pedestrian walkways, balconies, and vista points (ArchDaily 2016, INDER 2016, EDU 2014).

Parque Biblioteca León De Greiff

Part of the Medellin Public Library System (Sistema de Bibliotecas Públicas de Medellin), Library Park León de Greiff was designed to facilitate educational, cultural, and social transformation through neighborhood articulation. Located at the intersection of different neighborhoods, the library offers a green space, outdoor structures such as gyms and playgrounds, a city vista point, and pedestrian walkways to the central plazas of neighborhoods such as Villa Hermosa, San Miguel, La Ladera, Enciso, Boston, Los Ángeles, Batallón Girardot, Sucre, Prado, Los Mangos, Portal de Enciso, La Mansión, and Caicedo (SBPM [n.d.]).

Parque Recreacional Las Estancias

Las estancias is a recreational and sports infrastructure developed by the Urban Development Company (Empresa de Desarrollo Urbano or EDU) as part of the Center-east City Integration Project (Proyecto Urbano Integral, or PUI, Centrooriental). With an area of 14,950 m^2, the project sought the renovation and improvement of the existing Las Estancias sports field, replacing the old sand field with a synthetic one and adding grandstands for audiences (EDU 2013).

Camino De La Vida

The Camino de la Vida is one of two hiking paths leading to the eco-parks that constitute the Medellin Circus Garden (Jardín Circunvalar de Medellin). More than a mere tourist attraction, the Camino de la Vida also connects the majority of neighborhoods in Comuna 8 with the municipality's rural lands, towns, and exterior

[7] UVA stands for Unidad de Vida Articulada. UVAs are places for cultural development, recreation and amusement located in reconditioned water storage infrastructures built by the Empresa de Servicios Públicos de la Ciudad (The City Public Service Company). The UVAs offer programming for children, teens and seniors, as well as being public parks.

infrastructure projects. Built to generate public space for community use and enjoyment, the route consists of over 3.1 kilometers of hiking trails with associated green space. Various archeological sites were also recovered along the trail (EDU [n.d.]b).

Ruta De Campeones

La Ruta de Campeones is the second project that connects the Medellin Circus Garden to the city through Comuna 8. It has educational hiking paths, a bicycle path, and a path for individuals with limited mobility. The project was developed in three stages (EDU 2015).

These nine projects are clear examples of "Social Urbanism" policy, which invests in urban development in the most marginalized sectors of the city in an attempt to recompense the social debt owed to Medellin's poorest populations. Nonetheless, due to their largely architectural focus, these projects exhibit some significant failures when analyzed through the lens of external habitability, even though they might add to the aesthetic value of the territory.

Methods

The case study of the nine "Social Urbanism" projects described above was conducted in two phases. The first phase consisted of a comprehensive literature review of the existing information about the projects from sources found in online databases and physical archives. Legal requests for information[8] were sent to different municipal authorities to supplement the information and resolve incongruencies. Three documents formed the baseline of the secondary research. First was the municipal Territorial Organization Plan (Plan de Ordenamiento Territorial or POT).[9] The other principal sources of secondary information were the "Household Qualitative Deficiency Indicators" and the "Multidimensional Conditions of Life Index" produced by the municipality of Medellin. These sources provided quantitative data for geographic analysis about changes in land use caused by the intervention projects, changes in area of public space over time, intervention in protected lands, increase in area of green spaces, and the change in the number of residences exposed to unmitigable environmental risk over time.

The second phase of investigation included primary research conducted in the areas of influence of the selected intervention projects. Three different instruments were applied: a general survey, a focalized survey, and direct observation. A total of 318 general surveys, a statistically representative sample size for the comuna, were applied. The application was structured according to city blocks, which allowed for homogenous coverage in chosen neighborhoods and sectors. In contrast, the

[8] Article 23 of the Colombian Constitution of 1991 guarantees the citizens the right to "Derecho de Petición", in which they can make legal requests for the release of information from any level of the government for "matters of personal or general interest".

[9] A POT is urban planning tool that states objectives, strategies and plans of action in relation to urban development and specifies the guidelines and legal norms under which each project is carried out. While projects are specifically related to the physical development of a territory, they are required to take into consideration social and environmental impacts as well.

focalized surveys were applied between 8:00 a.m. and 8:00 p.m. in five of the nine intervention projects: eco-parks 13 de Noviembre and Las Tinajas, UVA Sol de Oriente, el Camino de la Vida, and la Ruta de Campeones. A total of 106 focalized surveys were applied.

Surveys were used to supplement the quantitative data for geographic analysis of the areas of influence (and caused us to question the means of measurement found in the secondary research). However, our interpretation of habitability is incomplete without also considering appropriation, or the way people use space. Therefore, the information about the physical development of the territory was put in dialogue with information about perception and use of the nine projects. This information was collected in the same generalized and focalized surveys, as well as through direct observation, and analyzed through an evaluation matrix developed by the researchers.

The matrix consists of a series of more than 60 different indicators that measure physical and social impacts of the projects. The indicators are grouped into four descriptive categories—physical habitability, environmental security, social construction of territory, and human security—showing the results, impacts, and perception of each project.

In the matrix, appropriation is defined as a numerical index that represented the unification of three perception indicators: "Frequency of Use" of the new public spaces generated by the projects (FRUE),[10] the "Level of Recognition" of newly constructed structures on part of the inhabitants (GRNE), and "Level of Use" of the spaces by inhabitants (GUIE). Each indicator was assigned a value of 1, 3, or 5 in accordance with a Saaty measurement scale. The value for the index "Level of Appropriation of Public Spaces" (NAEP) was obtained by taking the mean of the values assigned to each component.

Finally, direct observation was used for the collection of data and the calculation of the indicator "Level of Aesthetic Integration of Social Constructed Imagery". For this indicator, three spaces were chosen for the analysis: el Camino de la Vida, la Ruta de Campeones, and Ecoparque 13 de Noviembre. Photographs were taken of the murals that were painted as the spaces were constructed. These photographs were later analyzed for the level of integration in accordance with the evaluation matrix. A visual analysis of the artwork was used to characterize the institutional strategies in the transformation of public spaces from an aesthetic perspective.

In the following section, we discuss some of the findings that stand out when comparing the quantitative and qualitative data about the territorial changes in the areas of influence of the intervention projects when putting the secondary and primary research in dialogue according to the indicators used in the matrix.

Results

At this point, we present an analysis of four factors directly related to habitability in the case study; qualitative deficiencies in housing (which is analyzed from the perspective of internal habitability), public space, green space, and territorial

[10] Abbreviations reflect the Spanish names of the indicators.

sustainability. The last three are analyzed from the perspective of enriching the discussion of external habitability.

Qualitative Deficiencies in Housing

Most commonly, habitability is understood as a synonym for adequate or dignified housing, and its analysis has been limited to an architectural sense as the guarantor of physical habitability. The municipal administration of Medellin uses this model of habitability to describe "Qualitative Deficiencies in Housing" and to calculate "Index of the Multidimensional Conditions for Life" (ICMV). The first indicator consists only of an analysis of the interior of a residence, where factors such as building materials and overcrowding are evaluated. It does not take into consideration the coherence between the architectural solution and the expectations, needs or demands of the family group as habitability does when analyzed from the perspective of necessities.

Table 3.1. Qualitative deficiencies in housing: an analysis of residences in Comuna 8.

Deficiency	Percentage of Residences
Dangerous or unstable materials used in flooring	14.70%
Mitigable Overcrowding	100%
Houses without Potable Water	21%
Houses without Sewage Systems	16%
Houses without Electricity	7%
Houses without Garbage Collection	10.10%

Source: Medellín Como Vamos (2019).

As shown in the table above, the only external environmental factors considered in the analysis of the quality of the residence are related to the access to public services and basic sanitation. The analysis ignores the role played by exterior space as a compensation mechanism when the residence offers insufficient private space, providing only minimal areas that, in many cases, replicate conditions of overcrowding. In these cases, the outside environs constitute the stage for public life- for meeting up, recreation, and enjoyment of nature- and satisfy "the need to make social networks with the other people who inhabit the space" (Moreno Olmos 2008).

In contrast, the ICMV was developed by the municipal administration of Medellin to provide a multidimensional measurement of the living conditions of the citizens in objective and subjective aspects. The ICMV measured 15 dimensions: (1) the quality of the residence and the surrounding environs, (2) public services, (3) the natural environment, (4) schooling, (5) illiteracy, (6) mobility, (7) physical capital, (8) social participation, (9) liberty and security, (10) vulnerability, (11) health, (12) employment, (13) recreation, (14) perception of quality of life, and (15) per capita income.

Although the goal of the index is to offer an integrated view of the concept of "Quality of Life," some of the variables considered fall short. For example, when analyzing the "Quality of the Residence and the Surrounding Environs" the design only included variables such as the predominant building materials in the floors and walls, ignoring sociocultural, economic, and biophysical factors. The indicators don't include dimensions outside the physical-spatial, nor any variables outside of the residence itself, such as public spaces and recreational structures, or environmentally based variables sch as risk management, plant cover, or mobility infrastructures.

When comparing the analysis of these two indicators with the proposed theoretical framework, the insufficiencies become apparent: they only consider limited factors related to the internal conditions of habitability, excluding the integrated analysis of habitability from the external perspective, and ignoring the environmental factors or conditions that guarantee the satisfaction of necessities.

One of the biggest problems that arises from the lack of analysis that integrates environmental factors and social consequences is the priority given to commercial land use over residential use. However, cities need to strive for conditions that promote the mixed use of space to establish and maintain a healthily balanced and varied social composition. Simultaneous use of the same space in different capacities is a desirable condition for a city. However, this decision should be made in concert with the residential communities, as they are the people who will benefit or be harmed by development. In the case of Comuna 8, residential use in the areas where social urbanism projects were built dropped between 0.4% and 26.6 percent. Residential use in the area of influence of the project increased only in the case of UVA Santa

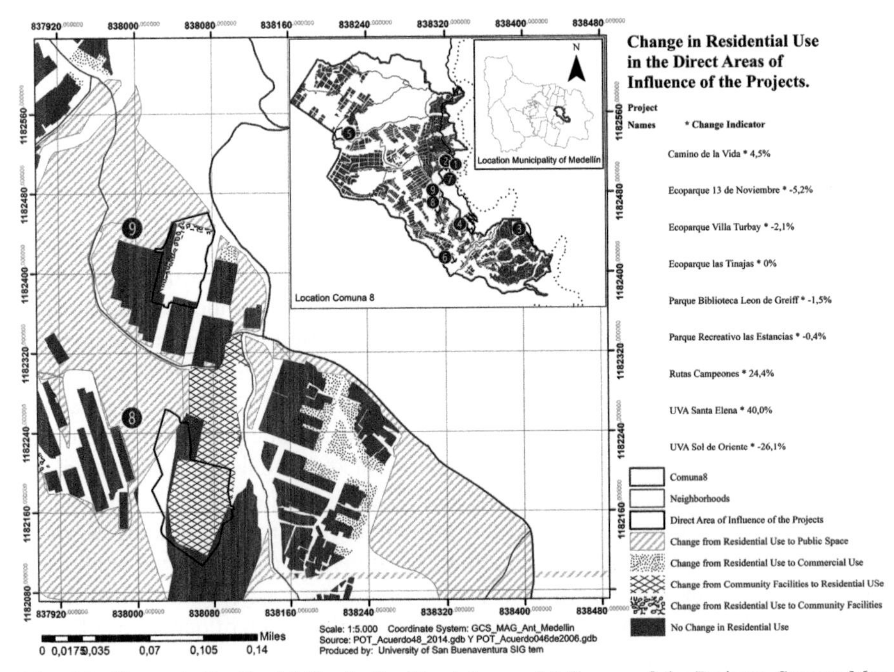

Map 3.2. Change in Residential Use in the Direct Areas of Influence of the Projects. Source: Map elaborated and produced by the researchers; indicators calculated based on the geographic information from Agreement 046 of 2006 and Agreement 048 of 2014.

Elena, growing by 40.03 percent. In general, the changes in land use caused by the urban intervention projects led to the displacement of local populations, promoting the arrival of new populations, linked to service and commercial industries.

Public Space

Public space is one of the most relevant variables in the analysis of external habitability because of the important role it plays in social interaction, the organization of public life, and social representation (Páramo et al. 2018). It should, therefore, be inhabitable and contribute to the overall quality of urban life. The focalized survey about level of appropriation of public space was applied in only five of the nine projects. In the majority of these, public space has a direct relation to social facilities; these facilities are given priority because they are understood to facilitate social interaction and community building.

"Social Urbanism" policies in Medellin are based on the supposition that the increase in area of available public space in square meters will inherently increase the appropriation and use of these spaces by the community. However, in the comparative analysis of the qualitative and quantitative indicators, we observed that in two of the projects, where there was a reduction in the area of public space, there was nonetheless a high appropriation of these spaces on the part of the community. For example, in the UVA Sol de Oriente, 47.4% of area dedicated to public space was lost as a result of the intervention project. Nonetheless, there was a high level of

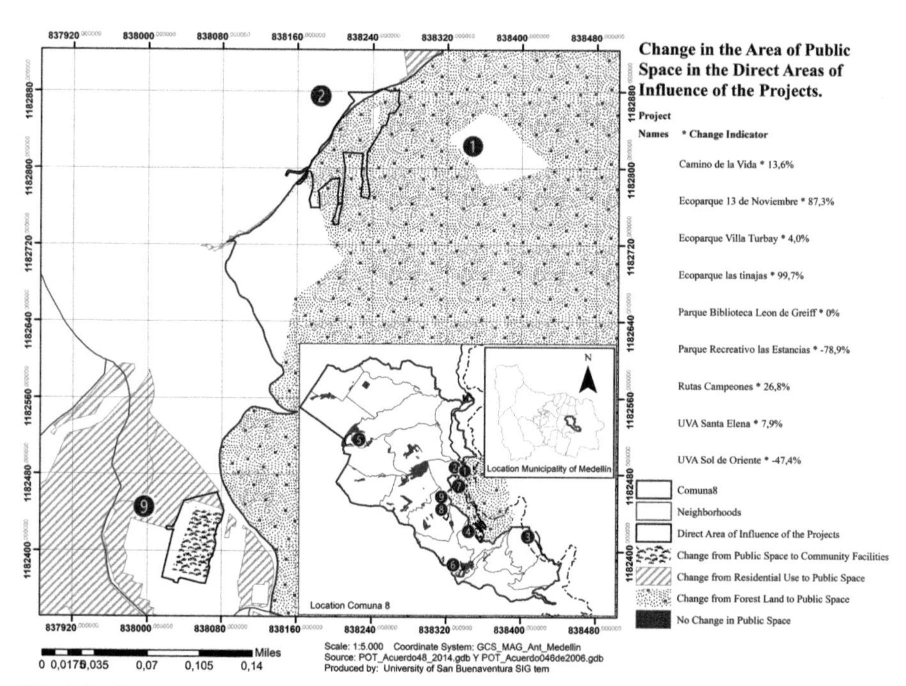

Map 3.3. Change in the Area of Public Space in the Direct Areas of Influence of the Projects. Source: Map elaborated and produced by the researchers; indicators calculated based on the geographic information from Agreements 046 of 2006 and 048 of 2014.

appropriation of these spaces, as reflected in the grade of usage (4.3) on the part of the inhabitants. A similar situation was present in Camino de la Vida, where the area of public space measured in square meters only increased by 13.6 percent. However, the consolidated index of appropriation, measured through surveys about perception, showed a high grade of appropriation (4.3) by the communities located in the direct area of influence of the project.

We only found one example that proved the institutional presupposition. In Ecoparque Las Tinajas, the area of public space, measured in squared meters, increased by 99.7% as a consequence of the urban intervention project. The resulting grade of appropriation by part of the community was high (3.7). The uniquely high score of this park can be explained by its intimate symbolic link with the neighborhood and the role of the space in local memory, as well as its close proximity to the neighborhood and the plant coverage that provides pleasant environmental conditions for spending time and meeting up with others.

Table 3.2. Level of appropriation of public space.

Project Name	Grade of Appropriation
Camino de la Vida	4.3
Ecoparque 13 de Noviembre	3
Ecoparque Villa Turbay	Not Applicable
Ecoparque las Tinajas	3.7
Parque Biblioteca Leon de Greiff	Not Applicable
Parque Recreativo las Estancias	Not Applicable
Rutas Campeones	3
UVA Santa Elena	Not Applicable
UVA Sol de Oriente	4.3

Source: Elaborated and produced by the researchers based on information gathered in perception surveys.

Green Space

Based on the indicator "Change in Area of Green Space in the Direct Areas of Influence of the Projects", we can afirm that 57.14% of the public recreational structures and social facilities analyzed neither lost nor gained additional plant coverage; the area of green space stayed the same. Nonetheless, plant coverage and green spaces in and of themselves are not guarantees of high levels of appropriation. They are only one of many factors that combine to make up external habitability, even though they do materially contribute to environmental comfort and create special microclimates which entice the community to inhabit and appropriate these spaces. An example of the disconnect between the presence of plant coverage and the appropriation of green space is seen in the UVA Sol de Oriente, where there was a significant loss of green space (50.9%), but nonetheless showed a high level of appropriation (4.3), which could be explained by other factors not considered in the indicator.

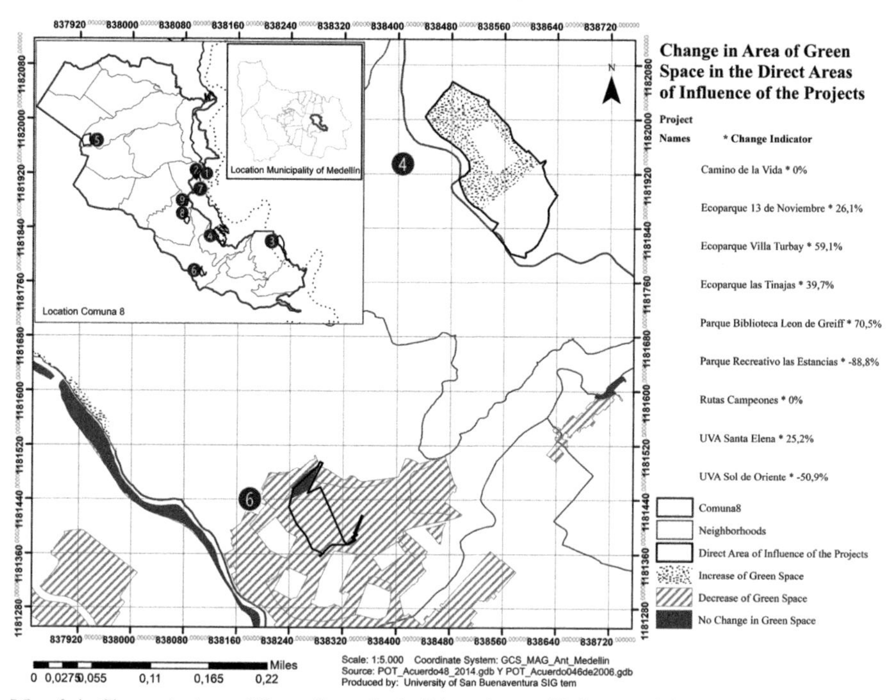

Map 3.4. Change in Area of Green Space in the Direct Areas of Influence of the Projects. Source: Map elaborated and produced by the researchers; indicators calculated based on the geographic information from Agreements 046 of 2006 and 048 of 2014.

After putting the data brought to light through the quantitative indicators in dialogue with the results of the perception surveys, we found that the increase in area of public space linked to the urban intervention projects does not necessarily guarantee a high level of appropriation, use, and recognition of the space. The significance of the space in the local memory of the community is a more likely guarantor of appropriation.

Territorial Sustainability

A sustainable theory of habitability should consider the resources used and the waste produced in the act of satisfying human necessities and in maintaining the ability to respond to those needs (Arcas-Abella et al. 2011, Casals-Tres et al. 2011). In Colombia, municipalities are required to seek territorial sustainability due to Law 388 of 1997. One of the strategies for the conservation of natural resources is declaring specific lands "protected lands", a status that excludes them from the processes of urbanization. Notwithstanding, the chosen projects show a disregard for the legal norm.

Comuna 8 is located on the eastern slope of the Valle de Aburrá, and is characterized by steep gradients and unstable lands, which implies a high level of environmental vulnerability and risk of landslides. External habitability should

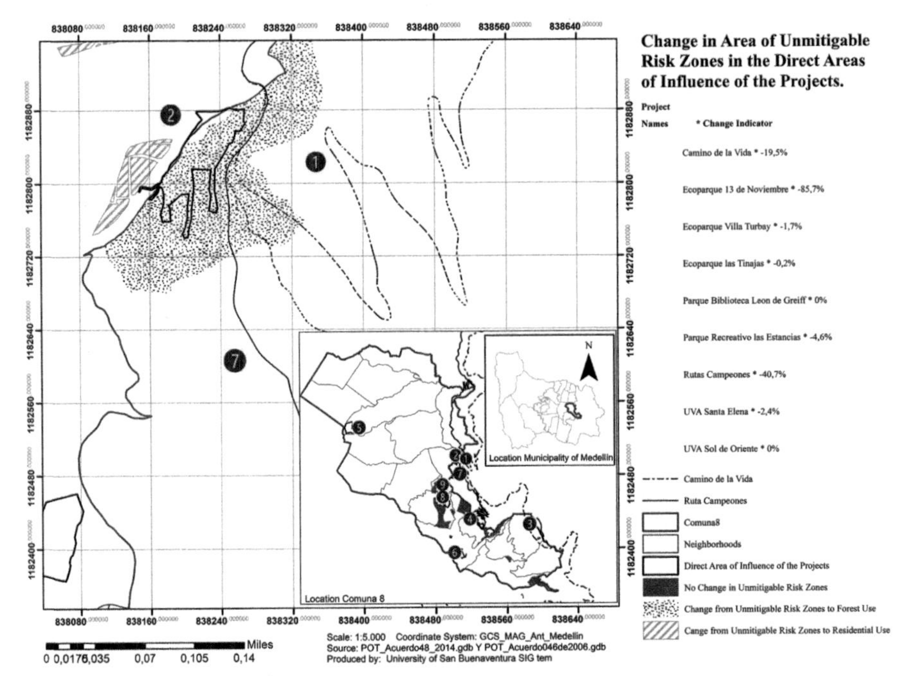

Map 3.5. Change in Area of Unmitigable Risk Zones in the Direct Areas of Influence of the Projects. Source: Map elaborated and produced by the researchers; indicators calculated based on the geographic information from Agreements 046 of 2006 and 048 of 2014.

consider physical security for the population that inhabits the slope; therefore, we included an evaluation of environmental risk and vulnerability.

We found a direct, but complicated, relationship between external habitability of the built environment and the natural environment. The urban intervention projects constructed to improve the built environment of the city have a high environmental impact that necessitates a new cost-benefit analysis.

For example, in the Territorial Sustainability indicator, seven of the nine intervention projects were analyzed. Of the seven, nealy half (44.4%) were built entirely in protected lands and 77.7% of the projects overlapped with protected lands by at least 70% intervening directly on shorelines and affecting riparian ecologies. This was the case for ecoparks Villa Turbay, Las Tinajas, 13 de Noviembre, and the UVA Sol de Oriente. These projects, which intervened in protected lands with negative environmental consequences, were carried out by the same municipal administration that lifted the zoning restrictions, allowing protected land to be used as public spaces or social facilities.

Ecoparque 13 de Noviembre is the most problematic case evaluated; built entirely on protected land, the project occupied 84.1% of the shores and creek beds in the area, with dramatic impacts on the riparian areas of local water sources. A similar situation was found in Ecoparque Las Tinajas, where occupation of the shorelines reached 54.5 percent. The interventions of both eco-parks negatively affected water resources and ecological structures planned for in the POT. Construction of the

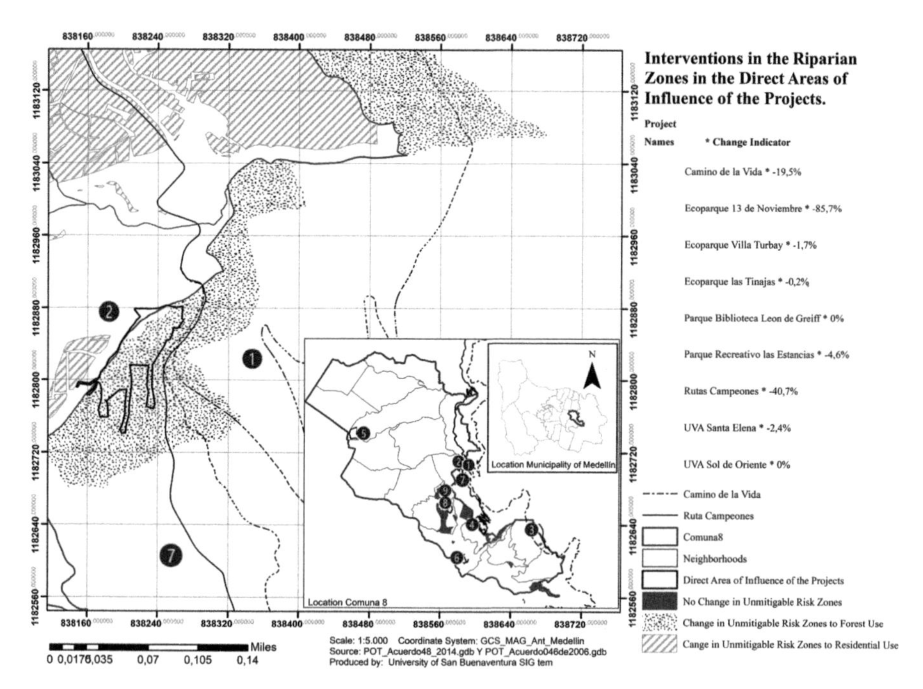

Map 3.6. Interventions in the Riparian Zones in the Direct Areas of Influence of the Projects. Source: Map elaborated and produced by the researchers; indicators calculated based on the geographic information from Agreements 046 of 2006 and 048 of 2014.

eco-parks was allowed because the municipal administration removed restrictions associated with protected lands reducing environmental vulnerability in the comuna, where the proportion of households living in environmental risk conditions reached 18.42% at its height. This decision allowed for the shorelines to be developed with city parks, in an effort to prevent squatting or the building of informal housing on shorelines. However, the municipal administration, in promoting the occupation of protected lands, sends a contradictory message to communities: protected land, which by law cannot be urbanized, can indeed be developed if the intervention is endorsed and carried out by the municipality.

In response to the urban intervention projects and the interventions along waterways, the perception of environmental vulnerability in the comuna has dropped considerably in 10 of the 16 neighborhoods where surveys were conducted; more than 41% of respondents affirmed that the level of environmental risk had decreased. Nonetheless, in stark contrast to the municipality's justifications for shoreline interventions and the improvement of perception of environmental vulnerability, the number of houses directly affected by landslides increased to 44.52 percent.

Conclusions

Understanding habitability from a perspective of necessities obligates the inclusion of the urban and territorial scales to understand their respective roles in the satisfaction of human necessities, the satisfaction of which does not only depend on the conditions of the residence alone, but also upon the conditions of its surrounding environs.

We understand that the results presented above account for the analysis of a limited number of indicators. However, we hope that with time, a consolidated index of habitability that integrates physical-environmental and social variables will be built, allowing for the full tracking of urban interventions in three scales: architectural, urban, and territorial. We recognize that the system can and should be adapted for analysis specific to the type of intervention to more clearly recognize its scope and limitations.

Despite the limited scope of the analysis, we find the results sufficiently compelling to offer three interventions to show the importance of expanding the concept of habitability to include both the internal and external analyses.

First, habitability from an internal perspective and based in the architectural scale alone does not allow for the integrated evaluation of the determinants of the quality of urban life for the populations located in the peripheries of the city. Additionally, an architectural analysis of dignity in housing does not take into account the social practices developed in response to housing conditions. For example, the practice of using public spaces when domestic space proves insufficient. The pitfall of defining habitability on an architectural scale is that it becomes impossible to talk about sustainable development for city growth and transformation, not to mention the failures of current measurements of internal habitability, such as the "Qualitative Deficiencies in Housing."

Our second intervention is related to public space. We understand public space not only as urban recreation infrastructures, but rather as places for real interaction and the configuration of social networks, where the recognition and appropriation of these spaces leads to a sense of identity and of belonging to the territory one inhabits.

Our third and final intervention speaks to sustainability. Like habitability, sustainability has often been understood only from the architectural scale, propagating the perception that adding to the built environment will decease environmental vulnerability. Interventions in the shorelines of the creeks and streams reflect this problematic view of sustainability. Through our study, we found that these intervention projects were built with direct interest in discouraging the construction of residences on the shorelines of waterways. Mitigating environmental vulnerability, however, was only part of the reason for these interventions, and the real interest was in controlling urban growth.

The three interventions discussed in this chapter show that habitability cannot be evaluated from the architectural scale alone and must also include analyses of the direct impacts of public infrastructures. It is necessary to redefine the concept of habitability, which on the one hand implies moving beyond the architectural level of analysis to reach the urban and territorial levels, and on the other hand, relates habitability to human needs to understand the satisfiers that go beyond the internal environment. Only by using an integral perspective of habitability, which puts internal and external habitability in dialogue, can we understand the impact of urban intervention projects on quality of life.

Acknowledgements

We recognize the support given by Helena Pérez-Garcés, Rector of the Engineering Faculty at the Universidad de San Buenaventura, Medellin campus, and her research assistant Maria Camila Gaviria-Peña, Environmental Engineer, in the process of georeferencing, consolidation of indices, and reviewing the text. We also thank Melissa Paucar Sánchez, Diana Valencia's research assistant, for her help consolidating the images and descriptions of the different intervention projects.

References

Arcas-Abella, J., Pagès-Ramon, A. and Casals-Tres, M. (2011). El Futuro del Hábitat: Repensando la Habitabilidad desde la Sostenibilidad. El caso español. Revista INVI 72(26): 65–93.

ArchDaily. (2016). UVA Sol de Oriente.

Alcalá Pallini, L. (2007). Dimensiones urbanas del problema habitacional. El caso de la ciudad de Resistencia, Argentina. Boletín del Instituto de la Vivienda (INVI) 22(59): 35–68.

Bentley, I. et al. (1985). Responsive Environments, A Manual for Designers. The Architectural Press.

Berhe, R.T., Martinez, J. and Verplanke, J. (2014). Adaptation and dissonance in quality of life: A case study in Mekelle, Ethiopia. Social Indicators Research 118(2): 535–554.

Bronfenbrenner, U. (1971). La ecología del desarrollo humano. Paidós, Barcelona, Spain.

Cámara de Comercio de Medellin para Antioquia. [n.d.]. Perfil socioeconómico de Medellin y el Valle de Aburrá. *In*: Informes en Estudios Económicos. Cámara. De Comercio, Medellin, Colombia.

Casals-Tres, M., Arcas-Abella, J. and Pagès-Ramon, A. (2011). Habitabilidad, un concepto en crisis. Sobre su redefinición orientada hacia la sostenibilidad. Informes de La Construcción 63(Extra): 21–32.

Castro, M.E. (1999). Habitabilidad, medio ambiente y ciudad. II Congreso Latinoamericano: El habitar. Una orientación para la investigación proyectual. México: Universidad Autónoma Metropolitana.

Cinturón Verde. [n.d]. Dos nuevos ecoparques se construyen para el Jardín Circunvalar. Medellin, Colombia.

Clarke, D.L. (ed.). (1977). Spatial Archaeology. Academic Press: London.

Craglia, M., Leontidou, L., Nuvolati, G. and Schweikart, J. (2004). Towards the development of quality of life indicators in the 'digital' city. Environment and Planning 31: 51–64.

DANE. (2019). Resultados Censo Nacional de Población y Vivienda 2018. Valle de Aburrá, Colombia.

Dasgupta, P. (1988). Trust as a commodity. *In*: Gambetta, D. (ed.). Trust, Making and Breaking Cooperative Relations. Blackwell, New York, USA.

Defensoría del Pueblo. (2004). Desplazamiento intraurbano como consecuencia del conflicto armado en las ciudades.

Diener, E. (1984). Subjective well-being. Psychological Bulletin 95: 542–575.

Drucker, P. (1983). The Concept of the Corporation. Routledge, New York, USA.

Echeverry, A. and Orsini, F. (2010). Informalidad y Urbanismo Social. *In*: Medellin Medio ambiente urbanismo y sociedad. Fondo Editorial Universidad Eafit, Medellin, Colombia.

Empresa de Desarrollo Urbana (EDU). [n.d.]a. Con el inicio del ecoparque Trece de Noviembre, obra del Jardín Circunvalar de Medellin, se crea más espacio público para la ciudad. Medellin, Colombia.

Empresa de Desarrollo Urbano (EDU). [n.d.]b. Visitas masivas al Camino de la Vida, obra del Jardín Circunvalar de Medellin. Medellin, Colombia.

Empresa de Desarrollo Urbana (EDU). (2013). Parque Deportivo y Recreativo Las Estancias. Medellin, Colombia.

Empresa de Desarrollo Urbana (EDU). (2014). Ecoparque Las Tinajas y Aula Ambiental Sol de Oriente, obras del Jardín Circunvalar que transforman la vida en la comuna 8. Medellin, Colombia.

Empresa de Desarrollo Urbana (EDU). (2015). Ruta de Campeones, un proyecto que se construye a muchas manos. Medellin, Colombia.

Empresas Públicas de Medellin (EPM). (2013). UVA Santa Elena. Medellin, Colombia.

Jimmy, E.N., Martinez, J. and Verplanke, J. (2019). Spatial Patterns of Residential Fragmentation and Quality of Life in Nairobi City, Kenya. Applied Research in Quality of Life.

Jeong, J.-H., Cheon, D.-Y. and Han, S.-H. (2019). A Better Maintenance Strategy, a More Sustainable Hanok: Towards Korean Traditional Public Facilities. Buildings 9(11).

Holcombe, R.G. (1995). Public Policy and the Quality of Life: Market Incentives Versus Government Planning. Greenwood Press.

Instituto de Deportes y Recreación (INDER). (2016). UVA Sol de Oriente. Medellin, Colombia.

Kelly, G.A. 1955. The Psychology of Personal Constructs, Volume 1: A Theory of Personality. New York: Routledge.

Landázuri, A. and Mercado, S. (2004). Algunos factores físicos y psicológicos relacionados con la habitabilidad interna de la vivienda. Medio Ambiente y Comportamiento Humano. 5(1-2): 89–113.

Levy, L. and Anderson, L. 1980. La tensión psico-social. Población, ambiente y calidad de vida. México: El manual moderno.

Martinez, J., Verplanke, J. and Miscione, G. (2016). A geographic and mixed methods approach to capture unequal quality-of-life conditions. pp. 455–472. *In*: Phillips, R. and Wong, C. (eds.). A Handbook of Community Well-being Research. Springer, Dordrecht.

Marquina, D. and Pasquali, C. (2006). Impacto del entorno en los niveles de satisfacción proporcionados por viviendas de interés social. Perfiles 26: 25–40.

Max-Neef, M., Elizalde, A. and Hopenhayn, M. (1986). Desarrollo a escala humana. Cepaur: Santiago de Chile, Chile.

Medellin Como Vamos. (2019). Informe de Calidad de Vida 2018.

Medina Gallego, C. (2001). ELN: Una historia de los orígenes. Rodríguez Quito Editores.

Mercado, S. and González, J. (1991). Evaluación psicosocial de la vivienda. Infonavit, México.

Mercado, S., Ortega, P., Estrada, C. and Luna, M. (1995). Habitabilidad de la vivienda urbana. UNAM, México.

Moreno Olmos, S. (2008). La habitabilidad urbana como condición de calidad de vida. Palapa 3(2): 47–54.

Morgan, C. and Murgatroyd, S. (1994). Total Quality Management in the Public Sector: An International Perspective. University Press, Buckingham, England.

Pacione, M. (2003). Urban environmental quality and human wellbeing—a social geographical perspective. Landscape and Urban Planning 65(1–2): 19–30.

Palacios, M. and Serrano, M. (2010). Colombia y México: las violencias del narcotráfico, en Seguridad nacional y seguridad interior. Tomo XV de la Colección "Los grandes problemas de México México, D.F. El Colegio de México.

Páramo et al. (2018). La habitabilidad del espacio público en las ciudades de América Latina. Avances en Psicología Latinoamericana. 36(2): 345–362.

Páramo, P. and Burbano, A. (2018). Convivencia ciudadana en ciudades latinoamericanas. Universidad Pedagógica Nacional, Bogotá, Colombia.

Pérez Maldonado, A. (1999). La construcción de indicadores Bio-Ecológicos para medir la calidad del ambiente natural urbano. Mérida: Facultad de Arquitectura y Arte de la Universidad de Los Andes.

Santos, L.D., Martins, I. and Brito, P. (2007). Measuring subjective quality of life: A survey to Porto's residents. Applied Research in Quality of Life 2(1): 51–64.

Senlier, N., Yildiz, R. and Aktas, E.D. (2008). A perception survey for the evaluation of urban quality of life in Kocaeli and a comparison of the life satisfaction with the European cities. Social Indicators Research 94(2): 213–226.

Shumi, S., Zuidgeest, M.H.P., Martinez, J.A., Efroymson, D. and van Maarseveen, M.F.A.M. (2015). Erratum: Understanding the relationship between walkability and quality-of-life of women garment Workers in Dhaka, Bangladesh. Applied Research in Quality of Life.

Sirgy, M.J., Gao, T. and Young, R.F. (2008). How does residents' satisfaction with community services influence quality of life (QOL) outcomes? Applied Research in Quality of Life 3(2): 81–105.

Sistema de Bibliotecas Públicas de Medellin (SBPM). [n.d.]. Parque Biblioteca León de Greiff – La Ladera. Medellin, Colombia.

Stanley, F. and Dornbusch, R. (1994). Macroeconomía. McGraw-Hill, Spain.

Tesfazghi, E.S., Martinez, J.A. and Verplanke, J.J. (2010). Variability of quality of life at small scales: Addis Ababa, Kirkos sub-city. Social Indicators Research 98(1): 73–88.

Tilly, C. (2003). Changing forms of inequality. Sociological Theory 21(1): 31–36.

Valencia, P. (2017). Marginalización urbana: entre la violencia y la paz. Análisis de sus equilibrios dinámicos: el caso de Medellín. *In*: Ciudadanías emergentes y transiciones en América Latina. Sello Editorial Universidad de Medellin-Universidad Autónoma del Estado de México, Medellin, Colombia.

Valencia, P. (2018). Los ciclos de la violencia ligada al narcotráfico en Colombia y México, una expresión de la ruptura de pactos con elites locales y nacionales. *In*: Natera, M y Valencia, P. (eds.). Políticas de seguridad y entornos violentos en Colombia y México. Universidad de Medellin, Medellin, Colombia.

Valladares, R., Chávez, M. and López, M. (2015). Elementos de la habitabilidad urbana. pp. 15–39. *In*: Valladares, R. (ed.). Diversas visiones de la habitabilidad. RNIU, Puebla, México.

Velásquez, F.E. (1988). Reforma urbana en Colombia. Medio Ambiente y Urbanización, 42.

Villagrán García, J. (1963). Teoría de la Arquitectura. Departamento de Arquitectura INBA. Cuadernos de Arquitectura, 13.

CHAPTER 4

Practicalities of a Cross-curricular Approach of Environmental Education

Andra-Dina Pană

Introduction

Cross-curricular activities offer a relevant approach in worldwide education nowadays. The cross-curricular approach addresses every school subject beyond the mono-disciplinary perspective, based on the concepts, notions, methodology, skills, and values which are shared by two or more sciences. The relationship between the school subjects/sciences involved in a lesson defines the type of the cross-curricular approach: multidisciplinary, interdisciplinary, or transdisciplinary (Ciolan 2003). How do we distinguish among these three different levels of cross-curricular approaches? There are three criteria: the content, the learning experiences, and the scope (Ciolan 2003).

Multidisciplinarity is defined in terms of shared knowledge, topic-based learning, and content-based connections between school subjects (Ciolan 2003). For example, a lesson called "Autumn" uses theoretical knowledge specific to geography and biology and creates connections between these two sciences: e.g., why seasons are formed and how the seasons change plants', animals', and humans' lives. Moreover, during a school trip or outing, as a non-formal educational activity, students can be taken to a farm or a forest to see what autumn really looks like. They can also be asked to draw, paint, write/recite a poem, sing/compose a song about autumn, or design or answer an autumn-related riddle. Each science involved makes use of its own theoretical framework (Halloun 2020).

Interdisciplinarity is defined in terms of shared skills, problem-based learning, and skill-based connections (Ciolan 2003). For example, a natural phenomenon which

Independent researcher, Bucharest, Romania; ghadarod@yahoo.com can be explained using physics, chemistry, mathematics. The sciences involved in an interdisciplinary relation are close, and come from the same realm (Halloun 2020).

Transdisciplinarity is defined in terms of values and attitudes, project-based learning, and solutions to real life problems (Ciolan 2003). For example, building a birds' house is a life problem which uses knowledge and skills from a wide variety of school subjects. It also teaches values and attitudes specific to teamwork and respect for nature. Transdisciplinarity leads to new sciences, as it offers solutions to everyday life problems and its outputs are creative and innovative (Halloun 2020).

Although interdisciplinarity is often mentioned in connection with environmental education (Palmer 1998, Vincent 2010, Holfelder 2019), I have shown above that both multidisciplinary learning activities and transdisciplinary based on projects and problem-solving tasks are useful when teaching environmental education. What kind of approaches are best for different types of environmental education activities suggested by Wu et al. (2019): lecture activities, exhibition venues and activities, community activities, outdoor leisure space activities? All these examples are open to any approach beyond monodisciplinarity. It is every teacher's decision where they lay the emphasis, either on the knowledge, on the skills, or on the values.

The purpose of this chapter is to frame environmental education as a subject to be part of a multi-, inter-, or transdisciplinary approach. The idea that is advocated here opposes teaching environmental education as a separate, mono-disciplinary school subject. The academic literature review strongly supports this approach. For example, according to Heimlich (1993), Palmer (1998), and Sobel (2014), environmental education is holistic in nature and application, based on its ability to permeate the whole curriculum both inside and outside school, based on shared basic concepts, shared skills, and orientation to real life problems. What this chapter intends to supply the practitioners of education with is how to plan and organize lessons in schools in order to teach environmental education blended with any other school subjects, ranging from arts to languages. Unlike Heimlich (1993), for whom environmental education involves moving in natural settings, it is defined by nonformal education. I intend to show that a regular classroom, too, can be the perfect setting for environmental education and not only while teaching knowledge, but especially skills and values.

This chapter starts from the definition of environmental education with the aim of understanding and presenting the concepts on which teachers build their class activities, such as theoretical knowledge, skills, values, learning objectives, assessment, students' age-specific issues. My research interest lies in teaching environmental education aimed at children because school environmental education should form behavior and conduct at early ages with the aim of forming an adult with environmental belief. I support my presentation with recent literature review resulting in annexes containing practical suggestions of lessons to be taught in schools based on the concepts discussed in the chapter.

Environmental Education Models

Within the universities in the United States of America, there are three different interdisciplinary environmental education models described by Vincent (2010): the

Systems Science approach/model (it emphasizes in-depth knowledge of the natural sciences and technical research and analysis centered on laboratory and fieldwork skills), the Policy and Governance approach/model (it emphasizes the social sciences, humanities, and public engagement skills, with a focus on public awareness and an emphasis on policy and governance processes which can lead to practices that can either threaten or create resilient and sustainable human-nature interfaces), and the Adaptive Management approach/model (it emphasizes coupled human-nature systems knowledge and both problem analysis and solutions implementation skills in often termed transdisciplinary processes). Vincent (2010) concludes on the profile of the professionals prepared in each of the programs presented above, so the students embracing the Adaptive Management model can serve as the meta-experts and decision process managers who understand the relevance of various expertise and knowledge claims in interdisciplinary and transdisciplinary processes and therefore can construct, facilitate, and manage these processes.

Although these models refer to the curricula in universities where environmental education is included as a bachelor or a master degree, they (will) also guide the decision makers who will decide the school curricula for environmental education or/and will become teachers for school children, therefore the university model will be transferred to school curricula. This understanding should make practitioners in education interested in analyzing university models, raising questions related to which model is appropriate to different ages, what school subjects are targeted, which skills are intended, which values and attitudes are encouraged, and which learning activities can be modelled. The wider the spectrum of learning activities the class provides, the more efficient environmental education is.

Environmental Education (Concept and Term)

A crucial question is related to what environmental education is. There is a wide range of answers and perceptions of people involved in education (policy makers, teachers, students, parents, Non-Governmental Organizations) who underline different aspects of the concept of environmental education, such as: theoretical knowledge, environmental awareness, environmental beliefs, recycling, and/or resources saving behaviors. Man is part of nature, which means that man and nature are not enemies, but partners. Therefore, this is the perception which this chapter totally opposes, as it is counterproductive from the life skill forming aim of education, is what Marques and Xavier's (2020) findings define as naturalist preconceptions of environmental education based on the man-nature dichotomy. A more useful approach is the systemic one, in which man is aware of the problems and responsible for both the human race and the planet's future.

In this chapter, I use the term environmental education not to define a school subject, but a cross-curricular approach in which to integrate as many school subjects as possible with the ultimate goal of forming the transferable skills which are intended by The Organization of Economic Co-operation and Development (OECD) for the year 2030 (cognitive and metacognitive skills, which include critical thinking, creative thinking, learning-to-learn, and self-regulation; social and emotional skills, which include empathy, self-efficacy, responsibility, and collaboration; and practical and physical skills, which include using new information and communication technology devices), in order for the current students to guide themselves in a complex and uncertain future world, as characterized by Howells (2018). The academic world is interested in defining the instructional systems that can effectively develop the knowledge, skills, and values specific to environmental education within the 2030 horizon.

As environmental education is one of the three priorities of OECD in terms of education for the future, this chapter tries not to offer full answers, but to make the topic-interested people question themselves and design efficient solutions to address environmental education within local or global contexts. In my opinion, environmental education should teach current students of all ages knowledge, skills, attitudes, and values which can help future adults survive the eco-changes. The curriculum of environmental education should comprise knowledge, skills, and values of many different school subjects (sciences, arts, humanities) within a cross-curricular framework.

Drăghici (2019) lists some of the sciences/disciplines/school subjects which support sustainable development: environment, biology, medicine, nutrition, agronomics, geography, engineering, architecture, citizenship, sociology, psychology, political science, history, law, economics, and business. Obviously, the wide variety of disciplines asks for a particular approach. As found by Wu et al. (2019), this approach requires changes in content and learning experiences. Innovative activities are required not only by the content, but by current students who belong to the generation of millennials', and who have a unique perspective on quality education, environment and technology.

I intend to provide examples of all academic subjects which help students gain knowledge, skills, values, and attitudes specific to environmental education. The emphasis will be on humanities, not on science subjects, as we need to extend the scope of environmental education from sciences such as physics, chemistry, biology (which already do environment-related teaching based on knowledge and skills) to subjects such as arts, languages, for example, which can also help teachers transfer knowledge and form skills, values, and attitudes in their students in the field of environmental education. The success of this approach depends on the school teachers' involvement (as a result of their own individual environmental awareness) or on the school curricula which are to be designed accordingly.

Environmental Literacy. Specific and Shared Knowledge, Skills and Values

The concept of environmental literacy is defined by different scientists in different ways, but the key concepts in all these definitions are: perception of and sensitivity to the overall environment, understanding and experience of environmental issues, environmental values and caring about the environment, skills for identifying and solving environmental problems, participation in solving environmental problems at all levels, responsible environmental behavior, conscious behavior based on personal responsibility and values aimed at avoiding or solving environmental problems (Vincent 2010, Howells 2018, Holfelder 2019). The concept was created by Roth (1992): "Environmental literacy refers to the comprehensive quality of knowledge, emotions, values, skills, and actions that people have about the environment and the human–environment relationship."

Peterson (2019) also speaks about ecological intelligence, ecological consciousness, and more importantly, ecological activism. In other terms, Peterson (2019) mentions the knowledge, the values and the skills, in this particular order, specific to environmental education. From this perspective, environmental activism means what students can do in relation to the environment in which they live, what their skills are in this field. The same applies for everyone in school education who is intended to become an environmentally-skilled adult.

Education for sustainable development "promotes competencies like critical thinking, imagining future scenarios and making decisions in a collaborative way" (UNESCO 2016). Different scholars rely environmental education on the development of theoretical and practical skills (systems-thinking competence, anticipatory competence, strategic competence, normative competence) (Aedo et al. 2017), motivational, volitional, and social readiness to solve problems responsibly in a variety of situations (Holfelder 2019).

Beyond the theoretical concepts which are specific to sciences and which are compulsory to be taught in school (such as habitat, food chain, protection, ecosystem, natural phenomena, resource depletion, species, water supply, living conditions, pollution level, preservation, nature hazards, ozone layer, recycling), environmental education aims at changing behaviors and new values and acquiring concepts, as proved by Marques and Xavier (2020). Other scientists, too, advocate for behavior change paradigm (Drăghici 2019) when it comes to environmental education.

There are several activities to be done while teaching environmental education. Some of them (learning theoretical knowledge, contemplating nature, for example) aim at knowledge transfer. The next ones generate skills (interacting with nature, solving problems, project implementation). Another one (asserting environmental ideology) creates values and environmental activism. From the methodological point of view, these activities must be presented to students gradually. The first step is exposing the students to nature, terminology and environment-related problems. Next, the students will be offered the chance to get involved with problems, so they will start forming environment-related skills. The result of acquiring knowledge and gaining new skills will lead to the students internalizing the values and creating environmentally correct attitudes and habits in an active life.

Drăghici (2019) presents the literature review on the competencies and abilities to be taught within the environmental education framework: systems thinking competency (the ability to recognize and understand relationships, to analyze complex systems, to perceive the ways in which systems are embedded within different domains and different scales, and to deal with uncertainty), anticipatory competency (the ability to understand and evaluate multiple futures—possible, probable, and desirable—and to create one's own visions for the future, to apply the precautionary principle, to assess the consequences of actions, and to deal with risks and changes), normative competency (the ability to understand and reflect on the norms and values that underlie one's actions and to negotiate sustainability values, principles, goals and targets, in a context of conflicts of interests and trade-offs, uncertain knowledge and contradictions), strategic competency (the ability to collectively develop and implement innovative actions that further sustainability at the local level and further afield), collaboration competency (the ability to learn from others; understand and respect the needs, perspectives, and actions of others-empathy; understand, relate to, and be sensitive to others—empathic leadership; deal with conflicts in a group; and facilitate collaborative and participatory problem-solving), critical thinking competency (the ability to question norms, practices, and opinions; reflect on one's own values, perceptions, and actions; and take a position in the sustainability discourse), self-awareness competency (the ability to reflect on one's own role in the local community and local or global society, continually evaluate and further motivate one's actions, and deal with one's feelings and desires), integrated problem-solving competency.

The reason why I have fully cited Drăghici (2019)'s categorization is because I want to emphasize that the skills above can be found in a lot of other school subjects: mathematics, physics, chemistry, arts, humanities. This reality is an extra argument in favor of the cross-curricular approach.

Examples of the values that environmental education intends to instill in students are: concern for environment, active participation, initiative, making choices (walk or drive, LED or light bulbs?) (Aada 2019), willingness to learn, waste avoidance (Maurer et al. 2020), perception of personal threat, self-efficacy, attitude towards recycling, attitude to norm, awareness of recycling benefit, (co-)responsibility, protecting the environment, respect for nature, closure to nature, humanistic altruism, biospheric altruism, openness to change, sense of civic responsibility (Popescu et al. 2020). Although their phrasing in the academic literature shows that they are specific to environmental education, in fact the values above are shared with other sciences and school subjects: geography, biology, arts, history, etc. Some of the values above are even transdisciplinary: e.g., responsibility, willingness to learn, self-efficacy, openness to change.

All the knowledge, skills, and values mentioned above are taught, transferred, or formed, in time, throughout school years and across the curriculum. Of course, like in the case of any school subject, formal education is to be supplemented by non-formal and informal education. Learning concepts, developing skills, adhering to values, following role models (teachers, parents, mates, etc.). What this chapter intends to emphasize is the efficacy of a cross-curricular approach of environmental education opposed to teaching environmental education as a new independent school subject.

In environmental education, teaching environment-related concepts are closely joined by skill-formation and practice learning experiences. Therefore, one may say that efficient environmental education is unlikely to happen indoors. No one denies the power of experience in environmental education, as described by Heimlich (1993) and Drăghici (2019), who suggest hands-on life experiences, such as site visits, internships and service learning in communities, project-based learning, place-based learning, field trip, and experience. According to Holfelder (2019), competencies are not based on any specific knowledge content, but rather more oriented towards questions of how the acquisition of the required competencies can be made possible. Thus, there comes the question related to the perfect setting in which environmental education can attain its aims. When organizing educative settings, the concept of nature-based education (Kuo et al. 2019) is proved to be useful, but not comprehensive. An innovative approach will include regular classrooms in the debate. Everything depends on the aims that every teacher decides to set for each lesson: cognitive, psychological, and affective lesson aims, based on Bloom's (1956) taxonomy revised by Anderson and Krathwohl (2001). Or, in terms of environmental education, knowledge transfer, skills formation, or values formation.

For example, the following formal and nonformal school activities have been observed, depicted, reviewed, researched, and assessed by Popescu et al. (2020): outdoor activities encouraging pro-environmental behavior, group projects for learners on environmental topics, digital games encouraging pro-environmental behavior, using various multi-media materials on environmental topics in class (video, audio, etc.), using virtual and augmented reality tools in class, online trainings addressing different level of education on environmental topics, initiation of pro-environmental activities for children and parents, introduction of after-school programs with classes about environmental protection and waste management topics, distribution of informational materials on physical support (written, audio, video), online distribution of environmental information, promotion of pro-environmental behavior in social media, special events on environmental topics (thematic days, guest speakers, conferences, etc.), internal and external contests on environmental topics, groups and communities for children and students focused on environmental topics.

As it can be seen, the list above contains varied activities in which different school subjects, places, venues, settings, generations, media, devices are combined. It is the wide range of options that allows each teacher and each student to find the appropriate type of activity to address their uniqueness in terms of ability, interest, level, etc.

It is an undeniable fact that environmental education is a life-long process (Palmer 1998). Therefore, academics need to set some guidelines which can help teachers organize the ages at which it is appropriate to transfer knowledge, to practice skills, to form values that are specific to environmental education.

The aim of environmental education is summarized by Wu et al. (2019): "citizens who receive environmental education will grasp a stronger awareness of environment, more skills of environmental protection, deeper consciousness of the relationship between humans and nature, thus joining in protecting the environment to promote sustainable urban development". Therefore, environmental education addresses both children and adults. It forms skills and teaches knowledge in order to create a generation to cherish nature and sustainable development and survive changes. The students' age makes a difference in which approach the teacher uses:

multi-, inter-, or transdisciplinary (Schreiber-Barsch and Mauch 2019) and on the content to be taught.

One vital difference in addressing environmental education to adults is that the strategy can start with creating what Rozman and Rozman (2020) name "a frame-of-mind", which can be translated into creating attitudes and values by making the adults adhere to the environmental values.

At young ages, too, environmental education starts with exposing preschoolers, for example, to practicing skills, such as recycling or saving electricity and water. Meantime, school should take advantage of the natural love of children for animals, which can become the opportunity for schools to design environmental education-specific activities which will be very effective as they start from strong emotions that are able to guarantee the knowledge transfer and skills.

The Learning Objectives of Environmental Education

I intend to phrase some learning objectives using Bloom's taxonomy (Bloom 1956, Anderson and Krathwohl 2001) which are attainable in a cross-curricular approach of teaching and learning environmental education. In order to exemplify, I randomly start from the following action verbs: *Remember* (Recognizing, Recalling); *Understand* (Interpreting, Exemplifying, Classifying, Summarizing, Inferring, Comparing, Explaining); *Apply* (Executing, Implementing); *Analyze* (Differentiating, Organizing, Attributing); *Evaluate* (Checking, Critiquing); *Create* (Generating, Planning, Producing).

In the annexes of this chapter there are examples of lessons that can be used in the cross-curricular approaches of environmental education. In each lesson sheet, I mention the school subject in the school timetable where the activities can be done, the age of the children, the objectives, the values, the skills, and the knowledge shared by the subjects involved and environmental education.

A. Geography lesson about natural disasters (Annex 1)
B. An art lesson about seasons. (Annex 2).
C. ESL (English as a second language) on adjectives (Annex 3).

Other examples come from the English textbook, *Impact*, National Geographic Learning series. The lesson planners suggested to teachers by the authors of the series discriminate between "content objectives" and "language objectives". The former ones define what the students will be able to do with the information presented in the reading, speaking, and listening activities presented in the unit beyond the new vocabulary and grammar structures specific to the target language. For example, "students will examine how their leisure-time activities impact the environment, analyze the environmental impact of bottled water and other plastics, identify ways to reduce, re-use, and recycle." (*Impact*, 4, Unit 4, Lesson Planner, p. 118 available online for the free use of teachers and students at https://eltngl.com/sites/impact/try-sample-unit/britishenglish?_ga=2.253751689.1874744219.1578860504-2103151505.1578860504).

Assessment in Environmental Education (The Learning-Teaching Process, The Environmental Literacy)

Assessment measures whether the learning objectives have been attained. The assessment or evaluation methodology depends on what the measurement is aimed at: knowledge, awareness, behavior, or the teaching-learning process. Tal (2005) speaks about a specific environmental education assessment framework including aspects of environmental knowledge as well as awareness, skills, values, and practical involvement, and addressing most of the environmental education activities' basic ideas and components. Popescu et al. (2020) support periodical evaluation of environmental knowledge and activities and underline that one important assessment criterion is organizing and participating in pro-environmental projects. Larijani (2010) considers initial evaluation of environmental awareness an important step to take for every student enrolling into environmental education.

Teachers make use of questionnaires and self-evaluation methods to evaluate the involvement and behavior of the students who have undergone environmental education. The assessment of environmental knowledge can be based on the same methods as students' science knowledge is measured. When it comes to assessing the environmental specific skills, specific methods are required. For example, recycling skills can range between "I recycle from time to time" and "I regularly recycle" for the habit of recycling.

What about assessing the environmental values that the students have adhered to after participating in cross-curricular environmental activities? Teachers need specific criteria in assessing the attained values of environmental education which is incompatible with standardized evaluation, as Wals and Van Der Leij (1997) prove. For example, according to Wals and Van Der Leij (1997), the evaluation should answer the question: "Who am I becoming?" not "What do I know?".

From my point of view, it is of utmost importance to assess the students' environmental behavior: whether they voluntarily and regularly take part in environmental activities. The supremacy of this assessment criterion derives from the fact that the students' behavior comprises their knowledge, their skills, and their values. In a cross-curricular approach in which environmental activities are organized, voluntary involvement in the lesson/activity/project is a clear indicator of the students' environmental values. Suggesting initiatives and starting environmental projects are indicators of higher performance in educational environment.

Teachers and education decision-makers also evaluate the learning-teaching process. In the case of environmental education, the paradigm suggested by Wals and Van Der Leij (1997) also applies. The criteria here are based on: the structure of the subjects (life-world related), the teacher's role (facilitator and co-learner), teaching strategies (real-world related and experiential), the student's role (participant, willing to change).

In order to assess how environmentally literate the students are, Roth (1992) defines three types of environmental literacy: nominal, functional, and operational, as follows: "Nominal environmental literacy indicates a person able to recognize many of the basic terms used in communicating about the environment and able to provide rough, if unsophisticated, working definitions of their meanings. Persons at the nominal level are developing an awareness and sensitivity towards the environment along with an attitude of respect for natural systems and concern for the nature and magnitude of human impacts on them. They also have a very rudimentary knowledge of how natural systems work and how human social systems interact with them. Functional environmental literacy indicates a person with a broader knowledge and understanding of the nature of and interactions between human social systems and other natural systems. They are aware and concerned about the negative interactions between these systems in terms of at least one or more issues and have developed the skills to analyze, synthesize, and evaluate information about them using primary and secondary sources. They evaluate a selected problem/issue on the basis of sound evidence and personal values and ethics. They communicate their findings and feelings to others. On issues of particular concern to them, they evidence a personal investment and motivation to work toward remediation using their knowledge of basic strategies for initiating and implementing social or technological change. Operational literacy indicates a person who has moved beyond functional literacy in both the breadth and depth of understandings and skills, and who routinely evaluates the impacts and consequences of actions; gathering and synthesizing pertinent information, choosing among alternatives, and advocating action positions and taking actions that work to sustain or enhance a healthy environment. Such people demonstrate a strong, ongoing sense of investment in and responsibility for preventing or remediating environmental degradation both personally and collectively, and are likely to be acting at several levels from local to global. In so doing, the characteristic habits of mind of the environmentally literate are well ingrained. They are routinely engaged in dealing with the world at large."

In accordance with the above model, the aim of assessing the environmental literacy is not to decide if the students pass or fail a course, but to see in which of the above categories they fall in order to choose the appropriate activities for them to advance to operational environmental literacy.

Since 2015, the Organization for Economic Co-operation and Development (OECD) has been interested in assessing environmental literacy using its Programme for International Student Assessment (PISA) (Paden 2012). The OECD intends to assess "how green the 15-year-olds are". At the same time, the conclusions of the evaluation process also refer to the design of the environmental courses in schools worldwide: "Across OECD countries, around one in five students is consistently able to identify, explain and apply scientific concepts related to a variety of environmental topics. In Canada, Finland and Japan, over a third of 15-year-olds have high levels of environmental literacy. Students acquire most information about environmental issues from school, although only a minority of students learns about these issues in stand-alone environmental science courses. Schools are a crucial source of information on environmental issues for students. While in most countries only a minority of schools has courses dedicated to the environment, the issue is often discussed as part of other core curricula, and many schools offer out-of-school activities that focus on the environment. Across countries, only a minority of students attends schools where learning about environmental issues takes place in standalone courses on environmental topics. Most students acquire their knowledge about environmental science in courses in related subjects, such as natural science or geography. When the subject is the environment, teaching and learning methods are often innovative. Increasingly, learning about the environment occurs outside the walls of the classroom. According to school principals, most 15-year-old students attend schools that use at least one out-of-classroom learning activity. Outdoor education is the activity most commonly reported, followed by trips to museums and science centres. On average, 77 percent of students in OECD countries attend schools that offer outdoor education, 75 percent are in schools that organise trips to museums, and 67% are in schools that conduct visits to science centres. However, the availability of such activities varies greatly across countries: in Japan, for example, 55 percent of students are in schools whose principal reported that none of these outside-school activities are made available to students, while in Portugal and the Slovak Republic, all schools offer at least one of these activities. Schools play a pivotal role in building knowledge about such crucial environmental issues as air pollution, energy, extinction of plants and animals, deforestation, water shortages and nuclear waste. Students most often learn about these subjects in school. Higher-performing students also use the media and the Internet to broaden and deepen their knowledge. By building students' skills in environmental science and referring to the environment throughout the curriculum, schools can help to foster an interest in the subject that persists beyond the school gates and into adulthood." (https://www.oecd.org/pisa/pisaproducts/pisainfocus/50150271.pdf).

Life Skills Taught by Environmental Education

The aim of environmental education is to prepare the future adults for the environmental challenges of the future: climate change, natural hazards, resources depletion, etc. For the people of tomorrow, facing the climate change and recycling will be life skills, as the World Health Organization (WHO 1994) has defined them, "the abilities for adaptive and positive behaviour that enable individuals to deal effectively with the demands and challenges of everyday life". There is not an exhaustive list of life skills unanimously accepted in the academic and pedagogical world, but the most cited and researched are: working in groups, understanding self, communicating, leadership, making decisions (Boyd et al. 1992), cooking, housework, using devices, solving conflicts, learning, time management, substance use, decision making, self-esteem, anger and anxiety control, dealing with media and advertising influences (Staab and Dick 2019), critical and creative thinking (WHO 1994), orientation in the job market, entrepreneurship.

Conclusions and Further Research

Although part of the academic world uses the concept of environmental education as an independent academic/school subject, this chapter pleads for a cross-curricular approach of the environmental topics (awareness, literacy, skills, and values) in school. By inserting environmental topics in all school subjects, their occurrence becomes higher, which will result in higher efficacy of the learning process. Environmental education can be part of every class and, moreover, it attains its educational aims and learning objectives when it is approached by the specific methods of different school subjects in order to create life skills and other environmental skills.

Beyond the theoretical models that this chapter presents, it has intended to suggest practical answers to teachers' and education policy-makers' questions related to the connection between different school subjects when approaching an environmental topic. The connections are skill-, content-, value-based, as all school subjects and environmental education share learning objectives. The annexes contain practicalities which are intended to make the topic-interested teachers and decision- and policy-makers find their own answers arguing the model lesson plans provided. The next step of my research is to conduct fieldwork in which the advocated approach is to be applied in classes. What I intend to do is to organize a focus group of teachers who accept to teach the activities in the annexes and then to conclude on the efficacy of such activities in education students.

The annexes offer three models of ordinary lessons of geography, foreign languages, and arts, in which the activities are designed to teach environmental education from a cross-curricular perspective at different ages.

Annexure

Annex 1

A geography lesson about natural disasters

Students' age: 16

Content Objectives:

At the end of the lesson, the students will be able to:

- exemplify natural disasters;
- explain the prevention and the safety measures to be taken;
- produce an Internet post on the topic;
- produce a piece of work in order to raise the awareness of other people on the topic;

Subjects involved in the cross-curricular approach:

Geography, Foreign languages, Art, ICT

Values: concern for environment, active participation, initiative, perception of personal threat, attitude to norm, (co-)responsibility, protecting the environment, respect for nature, closure to nature, humanistic altruism, biospheric altruism, openness to change, sense of civic responsibility.

Activities suggested:

1. Brainstorming

   Elicit words related to natural disasters!
2. Art session

   Illustrate a natural disaster by drawing, painting, or showing photos!
3. Brainstorming

   What can cause a natural disaster?
4. Brainstorming

   What can we do after a natural disaster?
5. Brainstorming

   What can we do to prevent a natural disaster from happening in our area?
6. Internet Search

   Search the instructions the authorities send to the population in case of a natural disaster.
7. Translate the instructions the authorities send to the population in case of a natural disaster in different languages (English, French, etc.).
8. Post it!

   Design a message about the prevention of natural disasters and post it online!

Annex 2

Art lesson about seasons

Students' age: 8

Content Objectives:

At the end of the lesson, the students will be able to:

- Recall time spent in nature;
- Critique the good and bad parts of each season;
- Interpret poems;
- Produce a piece of art in which to present their favorite season;

Subjects involved in the cross-curricular approach:

English, Art, Music

Values: concern for environment, active participation, initiative, making choices, civic (co-)responsibility, protecting the environment, respect for nature, closure to nature, humanistic altruism, biospheric altruism, openness to change.

Activities suggested:

1. Art

   Draw, paint or make a collage in which to present your favorite season.
2. English

   Read a poem in which a season is described.
3. Debate

   Which is the most beautiful season? (in groups of 4, each student will plea for one of the 4 seasons; the others will bring counter arguments—hurricanes, droughts, floodings, etc.)
4. Music

   Sing or play a song about each season.

Annex 3

ESL on adjectives

Students' Age: 12

Content Objectives:

At the end of the lesson, the students will be able to:

- Exemplify beautiful places around the world;
- Produce a leaflet;
- Summarize the information about their region;
- Check the information about a geographical point;
- Compare two sites;

Subjects involved in the cross-curricular approach:

English, Geography, Art, ICT, History

Values: concern for environment, active participation, initiative, willingness to learn, attitude to norm, awareness of recycling benefit, (co-)responsibility, protecting the environment, respect for nature, closure to nature, humanistic altruism, biospheric altruism, openness to change.

Activities suggested:

1. Photo gallery

   The students bring to school photos of different places around the world. The teacher asks the students to express what they would feel like if they were in the places in the photos: *frightened* in the middle of a hurricane, *ecstatic* on top of a mountain from where you can admire the valley, *frustrated* in front of a polluted river, for example.

2. Playing tourists

   Design a leaflet for the tourists in your area. Include: a map, photos of important sights, historical information. Post the leaflet on the school website.

3. Peer checking

   Every student will write a short text in which to compare two cities, two mountains, two oceans or two rivers. The compositions will include a lot of factual information (the number of inhabitants, the height of the mountains, the length and the width of the rivers, the salinity of the oceans, for example) which can be checked and a lot of comparatives and superlatives of different adjectives. Then, each student will give their piece of writing to a classmate who will check the geographical information and the language accuracy. Of course, the students will be allowed to make use of a source of information (books, Internet websites), both while writing and checking the texts.

References

Aada, K. (2019). How to Promote Education for Sustainable Development? Vision of the Educational Situation and Its Contribution to Sustainable Development. The Eurasia Proceedings of Educational and Social Sciences 15: 6–12.

Aedo, M.P., Peredo, M. and Schaeffer, C. (2017). From an essential being to an actor's becoming: political ecology transformational learning experiences in adult education. Environmental Education Research (in press).

Anderson, L.W. and Krathwohl, D.R. (2001). A taxonomy for learning, teaching, and assessing: a revision of Bloom's taxonomy of educational objectives. Longman, New York.

Bloom, B.S. (1956). Taxonomy of Educational Objectives. The Classification of Educational Goals, Handbook1: The Cognitive Domain, Longman, New York.

Boyd, B., Herring, D. and Briers, G. (1992). Developing life skills in youth. The Journal of Extension 30(4).

Ciolan, L. (2003). Dincolo de discipline. Humanitas, Bucureşti.

Drăghici, A. (2019). Education for Sustainable Development. MATEC Web Conf. Volume 290, 9th International Conference on Manufacturing Science and Education – MSE 2019 "Trends in New Industrial Revolution", 13004.

Halloun, I. (2020). Differential Convergence Education from Pluridisciplinarity to Transdisciplinarity. White paper. Jounieh, LB: H Institute, April.

Heimlich, J.E. (1993). Nonformal Environmental Education: Toward a Working Definition. The Environmental Outlook. ERIC/CSMEE Informational Bulletin, May Issue: 1–9.

Holfelder, A.-K. (2019). Towards a sustainable future with education. Sustainability Science 14(4): 943–952.

Howells, K. (2018). The future of education and skills: education 2030: the future we want (in press).

Kuo, M., Barnes, M. and Jordan, C. (2019). Do experiences with nature promote learning? Converging evidence of a cause-and-effect relationship. Front Psychol 10.

Larijani, M. (2010). Assessment of environmental awareness among higher primary school teachers. Journal of Human Ecology 31(2): 121–124.

Marques, R. and Xavier, C.R. (2020). The challenges and difficulties of teachers in the insertion and practice of environmental education in the school curriculum. International Journal on Social and Education Sciences 2(1): 49–56.

Maurer, M., Koulouris, P. and Bogner, F. (2020). Green Awareness in action-wow energy conservation action forces on environmental knowledge, values and behaviour in adolescents' school life. Sustainability 12(995).

OECD. (2012). How Green are Today's 15-year's olds? Pisa in Focus 4 (in press).

Paden, M. (2012). NAAEE releases Framework for assessing environmental literacy: being used in 2015 OECD assessment. Journal of Education for Sustainable Development 6(1): 17–19.

Palmer, J. (1998). Environmental Education in the 21st Century. Theory, Practice, Progress and Promise. Routledge, London and New York.

Peterson, Eric. (2019). Ecopedagogy: Learning How to Participate in Ecological Consciousness. CONSCIOUSNESS: Ideas and Research for the Twenty-First Century. 7(7).

Popescu, S., Rusu, D., Dragomir, M., Popescu, D. and Nedelcu, Ş. (2020). Competitive development tools in identifying efficient educational interventions for improving pro-environmental and recycling behavior. International Journal of Environmental Research and Public Health 17(156).

Roth, C.E. (1992). Environmental Literacy: Its Roots, Evolution and Directions in the 1990s. ERIC: Columbus, OH.

Rozman, T. and Rozman, M.F. (2020). Education for sustainability: learning methods and the current state in Slovenia (A preliminary study). International Journal of Smart Education and Urban Society 11(1): 41–63.

Schreiber-Barsch, S. and Mauch, W. (2019). Adult learning and education as a response to global challenges: Fostering agents of social transformation and sustainability. 65(4): 515–536.

Sobel, D. (2014). Place-based education: connecting classrooms and communities closing the achievement gap: The Seer report. The NAMTA Journal 39(1): 61–78.

Staab, E. and Dick, G. (2019). Life Skills Learning Among Middle School Students. Undergraduate Scholarly Showcase Proceedings 1(2).

Tal, T. (2005). Implementing multiple assessment modes in an interdisciplinary environmental education course. Journal of Environmental Education Research 11(5): 575–601.

UNESCO. (2016). UNESCO Official Website.

Vincent, S. (2010). Interdisciplinary Environmental Education: Elements of Field Identity and Curriculum Design. A research study conducted by the Council of Environmental Deans and Directors of the National Council for Science and the Environment (in press).

Wals, A.E.J. and Van Der Leij, T. (1997). Alternatives to national standards for environmental education: process-based quality assessment. Canadian Journal of Environmental Education 2(1): 7–28.

World Health Organization. Division of Mental Health. (1994). Life skills education for children and adolescents in schools. Pt. 1, Introduction to life skills for psychosocial competence. Pt. 2, Guidelines to facilitate the development and implementation of life skills programmes. 2nd rev. World Health Organization (in press).

Wu, E., Cheng, J.-Q. and Zhang, J.-B. (2019). Study on the environmental education demand and environmental literacy assessment of citizens in sustainable urban construction in Beijing. Sustainability 2020(12): 241.

CHAPTER 5

The Traditional Tax System before the Challenges of Circular Economy and Green Markets

Proposal of Selective Incentives

Reynier Limonta Montero

"What is the difference between a taxidermist and a tax collector? The taxidermist takes only your skin."

Mark Twain

Introduction

The traditional tax systems have a wide potential of externality's corrections directed to make possible the development of circular economy and green markets. Which circumstances make the impact point of taxes inside green business optimum? What kinds of elements are sufficient conditions to use the extra fiscal dimension of taxes? The last questions explain, in part, the main objective of this paper: support the selective incentives as favorable category for the development of circular economy and green markets, all of them starting on the tax system postulates. This allows finding the equilibria point for the implementation of an optimum theoretical proposal.

Selective incentives are an important category from Olsonian conceptual system. They become a powerful tool because they allow the explanation of the determination of reasons of some behaviors. In this specific case, they make easier the elimination of behaviors by potentials beneficiaries from extra fiscal measures that would prefer, without its influence, another behaviors different than wished since

Faculty of Law, Universidad de Oriente, Altos de Quintero s/n 90500. Santiago de Cuba, Cuba; rlimonta@uo.edu.cu the tax system. Thereby they could make possible the abandon of economic linearity model and the adoption of circular economy patterns. The selective incentives do not confine to their recognition as theoretical equivalent of the tax incentives.[1] Let us clearly explain the reasons. The Selective Incentives (SI) emerged as an answer to the elements that are conducive to the collective action inside the well-known "Theory of Collective Action Logic" formulated by Mancur Olson (Olson 1965).

They make easier the existence of collective action in its positive form, whereas inhibiting it when acting since negative form away. There was a positive correlation between positive selective incentives and the collective action. At this case would be found then not only in the benefits provided by tax rules another institutions or rules outside to the Tax System could be acting as such. While the tax incentives just have a promotion function, selective incentives have a coercive function join to the same promotion function. The coercive mechanism operates as a kind of penalty. SI are directed to explain and modify the behavior of the subjects, whilst the tax incentives create the legal framework it entails, *per se*, a pecuniary resignation for commendable goals. A considerable amount of literature has been published about this last feature (Velarde Aramayo and González García 1997, Graham and Smith 1999, La Rosa 2001, Graham and Rogers 2002, Rezzoagli 2006, Roca 2010, Bazza 2012, Klemm and Van Parys 2012, Martín Queralt, Lozano Serrano et al. 2015, Almunia, Guceri et al. 2020, Eichfelder, Jacob 2020). It is necessary here to clarify that both are not mutually exclusive, indeed the tax incentives form part of the Selective Incentives, as positive selective incentives. All things considered, it seems reasonable to assume that one is content inside another, and that is why the effects are substantially different. Perhaps, this would be the most important difference, from a theoretical-practical approach, between both categories.

This is a transcendental point for circular economy; this point makes possible the growth of economic actors as part of circular economy and green markets. When the tax system identifies the power of SI, it makes possible the juridical order work as a whole, aimed at achieving the wished behavior. That is to say, only with these kinds of tools circular economy and green markets may have a systemic legal approach.

On the other hand, the union between circular economy and green markets is possible only if tax system identifies clearly, the notion of selective incentives as well as the proposal of this paper.

The Theoretical Matter of Tax System

The tax system is the clearest result of the state development in a remarkable part of social life. In other words, the conversion of tax collection in a coherent system was possible thanks to the rationalization and resignification of taxes. There has been a long way since despotism and arbitrariness, until justice principles set have entailed the changes of the state, society, and citizenship. Consequently, taxes became an important tool in the public management of economy. Therefore, it is not possible to understand the complexity of the modern public intervention inside economics process without the tax dimension. This view is supported by Neumark, who writes

[1] Credit Taxes.

"Actually, the state uses the taxes as instruments of economic policy either through of practices of directions or intervention" (Neumark 1994).

Tax system assumes a sort of identity inside any kind of society, it reflects goals, dreams, history, and maybe the future of the social system itself. Essentially, this feature is possible find it if the sense and goal of any tax in isolation inside the system is observed.

Approaches of Extra Fiscality Phenomenon

The abandonment of the tax category as an end in itself is an evident truth. Overall, scientific consensus can be surmised around the extra fiscal component in all kinds of taxes (Casado Ollero 1985, Varona Alabern 2009, 2010, Espadafor 2011, Bengoechea 2014). As Griziotti argues, it "...cannot be ignored that the announcement or application of any kind of taxes or even with the preparation of any tax measure are triggered non-tax consequences, directly or not" (Ugalde 2019, Yacolca 2019). This sort of consequences is referred as non-tax effects from taxes. There are several possible explanations for this matter. First, the economic actors are constantly looking to bring down its costs, and the benefits of tax system could provide minorations, tax credits, tax bonus, etc., can become a way for that. If analysis is done of several experiences, the common denominator is the use of the ability to create, modify, and reevaluate the behavior of taxpayers since tax management. Well-known examples are China foreign investment development, the diminution of world sugar consumption proposed since WHO (Lombard and Koekemoer 2020), the regulation of alcohol market around the world, the creation of a framework for petroleum investment market, the risks coverture mechanisms and its influence in the commerce, taxes to prevent the flight of capital, etc. (Gordillo Pérez 2017, Boadway 2019, Griffith, O'Connell et al. 2019, Jia, Li 2019, Kyari 2019, Night and Bananuka 2019). All of them have as common category the use of the possibility of modifying or favoring and inhibiting economic behavior of the economic actors since the tax postulates. Last argument explains the impact point on the economic fabric from tax liability and that is right according to the famous Kepler paradox[2] (OBJ Giraldo-Bedoya and García OBJ). Thereby, achieving a modification in the behavior of the taxpayers is a direct result of extra fiscal component.

Once the welfare state has relapsed, the epistemic notion of the development of the taxation policies included the extra fiscal policies. Such has been its development that Varona showed that a minimum degree of extra-fiscality is inevitable, since the logical thing is that when providing for a tax, they seek various situations outside the mere collection and because they deserve greater legal protection. However, it

[2] In the process of Mars planet observation, Johannes Kepler had supposed that the orbit described by this planet was circular; however there was a difference between the measurement from empirical observations and their predictions. Mars did not appear in the horizon. Thus, Kepler had to modify the theorization. Mars orbit was not a perfect circle, more suitable figure, and however less elegant it seemed, Mars was described as an elliptical optic. This example served the science philosophy to explain the relationship between the probable and the facts. The Kepler paradox in this case serves to understand the inevitable feature of extra fiscality immanent to any tribute, regardless of its reason of origin and its essential core.

specifies that strictly extra-fiscal taxes are those that are conceived in their foundation and structure to achieve a non-fiscal purpose and in order to achieve it, their essential elements are formed in a specific structure. Then, it must be distinguished, among genuine collection taxes, but which contain extra fiscal elements from those whose foundation and structure are designed to achieve objectives other than collection (Varona Alabern 2009, Chow and Wang 2017, Yan 2017).

There is a considerable amount of literature buttressing the differences between a tax with non-collection finality-purely extra fiscality, and collection taxes with gleams of extra fiscality (Braithwaite and Reinhart 2019, Culquicondor 2019, Allayarov 2020, Burman and Slemrod 2020, Sarin, Summers et al. 2020, Tørsløv, Wier 2020). I believe that inside the conditions of modern public finance management and tax policies, it is very improbable to find a kind of pure tax, isolated, with only the purpose of collection. A systematic vision denies this vision, including income tax. Although it could be said that the budget principle of non-affectation is a clear demonstration of taxation for pure collection, the several special regimes, the exemptions, the minorations before personal circumstances, and the attenuation of economic ability with the theoretical construction of the contributory ability shows solidly the existence of extra fiscality as an indivisible part of the contemporary conception of the tax.

Indeed, Musgrave demonstrated that "… even when public treasury operations involve monetary flows of income and expenses, the basic problems are not financial matters. It does not deal with money, liquidity, or capital markets. Instead, they are problems of resource allocation, income distribution, full employment, price level stability and development" (Musgrave and Peacock 1967, Musgrave 1969, Musgrave 1980, Musgrave and Musgrave 1992). This brings us to understand the scope of the ramifications of this meta-category. However, is clear at this stage that Keynesian ideologies are the vortex of these kinds of approaches since public finances. From a neoliberal point of view, the contraction of state in colossal proportions make the minimization of tax public policies easier, moreover the extra fiscality became a minor political instrument of the government. For instance, there is a kind of "gravitational centrality" inside tax administration and it is the reduction of tax rates pending a mediate fulfillment of the purposes of tax policy by non-public actors.

One question that needs to be asked, however, is whether the courts, in a specific case, could make viable the differences between both kinds of tax goals. Several courts have based decisions on the borders of extra fiscality, considering constitutional goals related with the citizen rights. That is to say, extra fiscal goals have been invoked in contrast of collection focus. This can be illustrated briefly by the judgment of Argentinian Supreme Court, in 2006, recognizing as lawful the exemption to VAT for language teaching in the private sector (Sicuelo 2012). Thus, in the judgement the social importance of language teaching activity is preferred in the collection of VAT.[3] Another example of that is The European Court of Justice Judgment in the case of Hughes de Lasteyrie du Saillant. According to this case ruling, the Court turned the tide established with the Werner judicial decision (Justicia 1993, van Arendonk and Engelen 2005). Both rulings deal with right to freedom of movement in EU.

3 Value-Added Tax.

But, in the Werner decision, the legality of the difference in taxation treatment for nationals who registered earnings inside one state and live outside, and who live inside same state was established by the Court. In contrast to this, in the second case, the European Court ruled against Werner jurisprudence, considering the legal supremacy of the right to freedom of movement on the taxation goals pursued by the European states.

Although these results differ mutually, they reflect our exposition thesis in this paper. The European court adopted both decisions based on other goals different to the taxation goals, first in 1993 ruled to reinforcement of the EEC,[4] and then in 2004 ruled by[5] OBJ, such as freedom of movement.

The border line between collection and non-collection purposes of the taxes is a priority in order to mark off the limit of the tax power. Anyway, it is a very hard task to explain, as we have stated before, the tax system, with a separation between their influences on the taxpayer's behavior from taxation mission *per se*. However, it is possible generate a theoretical approach to understand the extra fiscality phenomena and the influence of selective incentives inside the tax system. This approach is very interesting because it provides grounds for understanding and explaining the relationship between both categories.

It can therefore be assumed that the relationship is based on common behavioral component. The extra fiscality pursues a specific behavior even when it only appears as a secondary effect of tax measure. SI explain the individual behavior, but in a sort of internal phase. That is to say, it explains how kinds of elements are sufficient conditions to modify any behavior before its factual expression. Therefore, SI could be outside of tax system components, and even then, have an influence on taxpayer's behavior. This particular point will be an important issue for future research. It can thus be suggested that the extra fiscality phenomenon can be understood from SI theory, having as connection point the tax system.

At this point, focusing the analysis on the specific dimension of extra fiscality as an environmental tool, we would also like to emphasize that here are shown the fully developed effects of extra fiscality phenomenon. By way of illustration, one of the most relevant examples is the report presented by a group of experts in New Zealand called "The future of Tax". In that document they arrived at the conclusion that the "taxation is not simply a means of raising revenue. Tax can also be used as an instrument to achieve specific policy goals by influencing behavior. The group considers that the tax system can play an expanded role in New Zealand's environmental policy, helping to change behaviors and fund the transition towards a more regenerative, circular economy" (Barrett and Makale 2019). In connection with the above, the environmental taxes are the explicit renunciation of taxation goals in the interest of achieving better behaviors with nature. In the specific case of the circular economy and green markets, traditional tax systems react as any other genuine objective to protect and develop since incentives. But the point defended in this paper is relatively different—the green markets and circular economy need a special adequacy inside tax system. That is to say, a theoretical restructuring must be

[4] European Economic Commission.

[5] European Union.

made. According to OECD data, the average tax structure around the world consists of a kind of tetragon, in order of decreasing collection, goods and service taxes (including customs), other kind of taxes, individuals' taxes, and corporate taxes. On the basis of progressivity and taxation range, the extra fiscality in the case of circular economy and green markets should be expressed in a consistent treatment through tax tetragon. "Consistent treatment" should be understood as promoted behaviors with a tax reward, and it must not be confined to the credit taxes and should be directed to the internal behavior, which is exactly how SI works. Although the extra fiscality continues being the distinctive note around the tax phenomenon (Li and Zhu 2017), the new epicenter of any tax system which attempt to deals with circular economy and green market matters will be a structurally "consistent treatment". We'll illustrate this point later.

The Topic of Impact Point Modulation. Regularities for the Circular Economy

Circular economy supposes a closed-loop operation focus and a systemic change related to the resilience in the long term; it generates commercial and economic opportunities correlating the profits with the environmental goals. This economic system is in contrast to the traditional linear economy, which has a take, make, and dispose model of production. The circular economy is often referred to as circularity and implies an economic system that avoids waste and encourages the continual use of economic resources (Parchomenko, Nelen 2019, Choi, Taleizadeh 2020, Ghisellini and Ulgiati 2020, Kuah and Wang 2020, Makarova, Shubenkova 2020, Santagata, Zucaro et al. 2020). The last feature is located in the most outstanding point for the tax system: the correction of the linearity in economical process. In Figure 5.1, it is possible to warn about the points of importance for the tax system, not only as a supra objective anchored in the notion of sovereignty that is embodied by the tax power (Limonta Montero 2019), but as categories closely related to the essences of the tax system.

When Ken Webster designed the critical points of circular economy, such as design without wastes, increasing the resilience, the use of renewable sources of energy, the system thinking, falls thinking, and sustainable development (Pollard

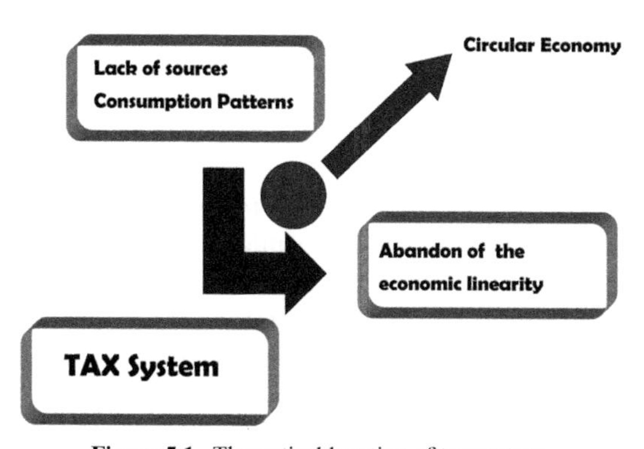

Figure 5.1. Theoretical location of tax system.

2016), a kind of roadmap for the rethinking of traditional tax system was being established. This point is valid equally for green markets particularly, and the green costumer phenomenon and its relationship with the global value chains. The observed correlation between critical points of circular economy and tax system might be explained under "consistent treatment" scopes. All critical points are meta-goals of the tax system and they must be treated as such. It is likely therefore, that there is a reflex of them in the tax collection. That is to say, the individual's taxes, corporate taxes, good and services taxes, and other kinds of taxes should express the selective incentives directed to promote the circular economy and green markets through the critical points. In other words, the critical points of circular economy should be a tax system wished behavior, and therefore must award the taxpayers who make reality. In contrast, the traditional behaviors must be punished. This penalty should be carefully designed and must, as general rule, face up the traditional tax rates and ordinary tax collection process. The impacts of the taxes on the economy have several dimensions. That's why the design phase is so important, in order to avoid undesirable effects. Considering the complexity and interconnection of the economic process, these are not only remarkable for the national taxpayers but to other states and, consequently, to other economies and taxpayers. This is not a minor phenomenon and is extremely relevant for the tax approach of the circular economy and green markets. One question that needs to be asked is the refined design of selective incentives being careful does not cause harm to other states or societies. Explaining the influence of tax designs beyond their borders, Baker and Murphy have pointed out "that identifying and discouraging policies and procedures that can cause harm to other states as part of an effective international moral harm convention is desirable"(Baker and Murphy 2019).

That is the reason by which a global approach, from international law, for the circular economy is absolutely desirable. However, from the purely scientific design, there are some common keys to signify the challenges of general ways for the tax system.

First, related with one of the most transcendent point of circular economy and green markets, the adoption of reward paradigm directed to productive matrix of recycling and reusing is very important. The minorations of tax rates in order to prioritize the investment process in green technologies and the renewable sources of energy could be successful if the collection process would have a positive selective incentive as part of "consistent treatment".

Although the conception of selective incentives was created to explain the logic of collective action (Olson 1965), its modulation on circular economy and tax systems relationship allows to locate the behavioral changes in two different scenarios. First, the conviction inside the taxpayers is directed to obtain results and profits from tax benefits, and subsequently develops the postulates of circular economy and green markets in their productive and distribution chains. These behavioral changes are produced in order to guarantee the success of the taxpayers' economical activities. It has the principal aim to mobilize all the members of the economic fabric until these goals, from the most relevant members to the least committed are identified with the circular economy goals and green markets (Vona, Marin 2019).

Second, the behavioral changes made easier by the governance tools as the tax benefits in a reward focus since tax system, lead to behavioral changes of action to guarantee the success of the economic activities, since it is not a simple externality, but a sure way to increase the competitiveness venture, since the management structures of taxpayers should welcome it as their own. In this same string of analysis, there are theoretical positions that support, rightly, that the environment is not a simple externality as long as it represents risk of all natural systems. Consequently, as we demonstrated before, the special adaptation of green markets and circular economy inside tax system is urgent. There are environmental externalities with a negative and positive sense, for example, the pollution of atmosphere, waters, soils, seas, but the classic externality is understood as a classic negative externality. However, its classic conceptualization explains a cost or benefit that falls upon an unrelated third party.

As the New Zealand tax experts report have explained, "*one example of a negative environmental externality is air pollution from an industrial plant that reduces air quality in a neighboring district: the residents in that district may have no connection to the industrial plant but nevertheless suffer the effects of the downwind air pollution. Externalities can also be positive. For example, a restored wetland might provide flood protection for the surrounding area and also improve water quality. Social and ecological enterprises also produce positive externalities*" (Barrett and Makale 2019).

The operation of the selective incentives that are simply embodied in advantages make the negotiation path towards the abandon of the economic linearity easier, modulating it according to the circumstances that the scenario described by tax framework provides.

The failure of linearity mode of economic activities become in a selective incentive to the actors under the tax postulates of reward focus and consistent treatment. It allows obtainment of profits under surface of the market, changing the way of realizing business, specifically adopting a new way to make business.

As explained above in this paper, the tax impact point in the circular economy and green markets are located in the development of taxpayer's behavior, characterized by the adoption of selective incentives as guidelines, since tax postulates framework to modulate the meta-goal of circular economy.

There are eloquent examples of this China inspired by the Japanese and German recycling economy laws, creating the concept of circular economy to reduce the environmental impact of the colossal manufacturing development of the Asian giant. Within this strategy, there are three very clear levels: at the level of the individual company, managers must seek efficiency through three basic principles—reducing the consumption of pollutants and waste emissions and resources, reusing resources, and recycling by-products. In the second level, within the eco-industrial parks and group industries, it is necessary to reuse and recycle the resources that are recycled within the local production systems (Du and Deng 2017, Tomasetta, Zucchella 2017, GOEL 2018, Brimsøe and Østerhus 2019, Cavaleiro de Ferreira and Fuso-Nerini 2019).

Finally, they have designed a third level of integration, in which the productive and consumption systems of the different regions and sub-regions of China are combined, and as a requirement, there is a palpable development at the local level of systems of collection and storage, processing and distribution of each product. In a global way, the precise system of cleaner companies that make their transition to the circular economy are owed partly to public facilities granted by the public administration.

This kind of facilities act as positive selective incentives, and later we will explain the role of the tax system in this pioneering legal framework.

In summary, the point of tax impact and the regularities of the circular economy would be graphed as shown in Figure 5.2. This relationship marked by the reward approach allows us to outline the epistemological line that defines the challenges for the circular economy before the tax system, as we will address in the following subheading.

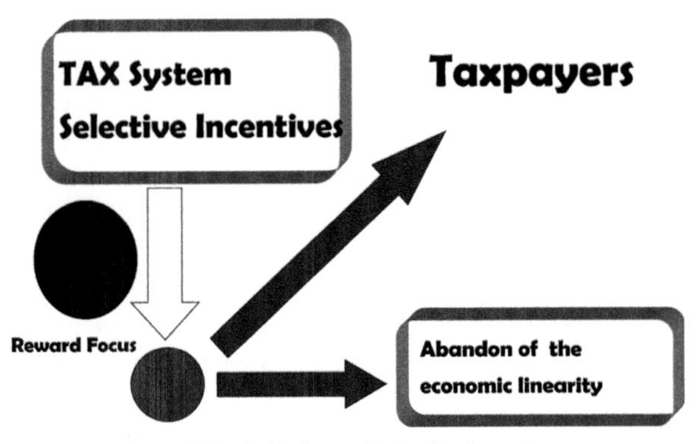

Figure 5.2. Guidelines of selective incentives.

The Tax System before the Challenges of Circular Economy and Green Markets. A Sort of Fine-Tuning

Just as Keynes defined the fine-tuning in the economy management (Cairncross 1978, Fontana, Realfonzo 2020), the relationship between the tax system and the circular economy must behave in the same manner and degree.

As we have explained before, the Chinese experience in the relationship between the management of the tax system, the green economy, and the green markets have some references in the Japanese and German legal framework. Their tri-dimensional characteristic is one of the most outstanding points present in the Chinese model. Those three levels are structured the following manner: a legal framework for the sound cycle society, energetic efficiency law, and Integral Planning of recycle defense.

In that same measure, the tax system from a reward focus based on "consistent treatment" should propose help on the economic subjects that make the transition between the conventional economy and the circular economy. In the case of the development of the green markets, I believe that the cascading tax or goods and service taxes should limit its impact on the consumer price of the products, making the competition easier.

Today, this is not the dominant focus inside comparative law. It is assumed that the green consumer belongs to a market segment willing to pay more for products with these features. In fact, it is a stereotyping approach. Numerous segments of consumers are excluded. Moreover, the power of the production, distribution, exchange, and consumption sectors related to the circular economy, in fact, would exhibit a palpable concentration that would hamper its maximum reproduction.

Although the tax approach to these challenges is desirable from a special tax regime, it is different from the general tax regime not only in tax rates, but in different management mechanisms that make their change easier, for instance, in tax deductible expenses inside the process of tax base establishment.

This could help to fight the increase on prices in the wholesale sector, which would directly influence the retail sector and competitiveness. Productive processes with little development in the circular or green approach could benefit from the tax influence. The last argument expresses a kind of "consistent treatment". We could take green logistics as a good example. This entails the use of better methods for attracting and handling raw materials, as well as Kümmerer's vision in the business model when designing minimizing entopic losses and finding in recycling an opportunity to obtain greater economic benefits. The fine-tuning that has been established becomes a reality in the application of the Selective Incentive Theory by combining the use of traditional methods as activities less weighted by the tax system and therefore with a greater objective or procedural burden. In fact, globally, the traditional economy can become a negative selective incentive. This, from the behavioral point of view, will favor stepwise migration towards the legal framework of the circular economy. The change in tax management and the taxation of this phenomenon will have as a common denominator the reward approach with positive and negative selection. Of course, the effect of the incentive will be variable, because it will depend on the impact on the business model in question. Matrices are never a uniform or simple phenomenon, and the theorizing we propose is a consistent treatment since tax system.

Otherwise, the model will not work because the ability to modify the economic management of taxpayers would be very low if there is no effective tax system. Then, cheating would become a viable strategy. The relationship with the types of selective incentives to be used cannot be a system of unique components, since the dynamism of economic phenomena would prevent it. It is, first of all, a circumstantial phenomenon, although its approach Is from a special regime and in a lane of instrumentation of three parallel tracks, which would be potentially desirable.

We have argued that the approach should be comprehensive, from an International Framework Convention, which from international law would allow starting its instrumentation at asynchronous speeds based on the real possibilities of each economy and legal system. The main elements of the model would be the reward approach and a consistent treatment from the tax system.

As it has been well argued in the New Zealand report, supra cited, it is not an externality. The change of productive and consumption patterns is today a necessity for the human species to survive the climate crisis, and therefore the adaptive advantages of our species: culture and society and in it the tax system included must change to contribute to that supreme goal.

A Model of Optimum Tax System for Circular Economy and Green Markets

Of course, to describe in a model the challenges of promoting a given productive approach is a hard task. However, I believe that four structuring elements can synthesize the challenge approach proposal. The model would consist of four primary structuring elements. As shown in the Figure 5.3:

a) Reward focus
b) New management
c) Green logistics
d) Selective incentives

They are not exclusive of other elements that could shape the sort of goal or critical sheet so that traditional tax systems approach the problem of changing production and consumption patterns.

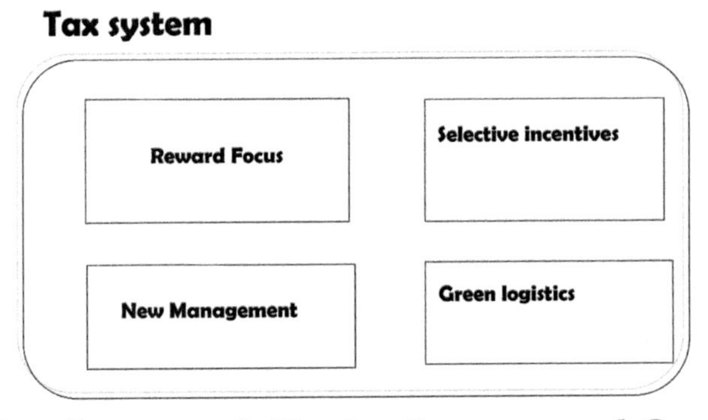

Figure 5.3. Primary structuring elements.

We have explained in detail the reward approach based on its importance in the fine-tuning management, however we will systematize them by explaining that its embodied is not a fiscal sacrifice, but that the effect on the development of a circular productive matrix corresponds to a kind of compensation for indirect amounts that in the fiscal exercise of early stages must suppose the reduction or weighting of the tax rates. Further studies, which take these variables into account, will need to be undertaken. The reward approach is the key to start the transition of productive matrix with other integral parts of selective incentives, such as public development policies, local decentralization, and environmental education and awareness, that are not the subject of this chapter.

The new tax management has a procedural function, because the tax levy process can be discouraging in itself. The establishment of practical, simple, and favorable self-assessment rules for the components of the circular economy would be desirable. In this particular case, the three levels already established by various legal systems could be an interesting guide in tax management. When recycling benefits are established, there is a reduction of the tax pressure on industries that find in recycling a part of the raw materials to be submitted to the manufacturing process. A monetary stimulus is actually carried out, since it avoids the transit of capital that can be returned to the expanded reproduction of the given business, and secondly favors responsible behavior, which has an appreciable behavioral-educational effect for the economy of the subjects.

On the other hand, the new approach must also relate cross-border business and, of course, the customs system. This is why we defend the necessary global approach on this matter.

Green logistics is an important trend because it allows the production and commercialization cycle to be closed and, therefore, the ecological footprint comprises all the stages of social production. Now the phenomena that it comprises can not only be seen from the reduced variant of utilization of renewable energies, which have a high ecological footprint in many cases. It is the adoption of efficient supply practices that modify the harmful management environment at all costs that is observed in large distributors whose logic is focused on profit and market factors. It is an evident truth that the tax system is only one of the influences in this change of matrix, for logistics in particular, but within it has negative selective incentives. In this case, when fulfilling the function of the sanction, they should put worse off those who refuse to initiate the transition of productive matrix, which will keep two very clear positions. One in which the freedom to choose technologies is guaranteed only when the cost and competitiveness of the business should be feasible under an ecologically directed tax pressure, and secondly that the benefits will privilege the least damage to the common good in the productive processes. These approaches are not simple; in fact, they have a deep philosophical anchor, which guarantee the sign of contemporaneity in the production and provision of services.

We have argued that selective incentives are a theorization with an elegant application in the situation described. We have approached his explanation from the positivity, so we will focus on the negatives in this case.

The latter's must be understood with a teleological approach and the operational mechanism is very similar to that of the sanction in the field of the theory of law. The first assumption in relation to incentives states that when the costs of individual contributions of collective action are reduced, there is a clear tendency for the individual to discourage even to investigate whether it is worth it or not to make a contribution, even in the intuitive level it works that way. However, the question of incentives does not remain as a *clausus* number in the Olson conceptual apparatus, as it clarifies that if the group that will take advantage of the collective action, and it is sufficiently small and the cost-benefit ratio of this action is sufficiently favorable, it can be an action calculated in collective interest although there are no selective incentives. These theoretical approaches serve as a kind of foundation of interaction between actors operating with circular economy technics and its potential partners inside economic and change relationships. It allows the understanding of the possible-rational elections in the field of decisions on circular economy and green markets.

Before, we explained based on the advantages that the use of circular economy phases is incorporated in the governance of the business.

The interaction of these four configurator elements is complex, and on a rationality basis. We have already argued that the univocal component thesis is neither realistic nor practical, given the asymmetries of the economies and therefore of the characteristics and functional logic of the tax systems. It is an epistemic guide, a tetragon that marks a corrective function for the challenges of resilience and circular economy as a peculiar form of management in economic processes. This model operates from a presupposition of functionality of the pre-existing tax system as any previous not functional would make the propositions made in this paper unfeasible.

The proposed model, without being a self-evident phenomenon, is not a rigid panacea, in fact its shaping elements are extremely flexible to accommodate the probable changes in the economy and the global environment. It is a vision to begin to transform the vision from the tax law in relation to the new management models that have been regulated in the tax matter, but with a biased and insufficient approach.

Starting to change is difficult, breaking schemes is even more difficult, in the case of financial and tax law in particular it will mean the reconfiguration of its systems of principles, operational guarantees and basic institutions, since everything was built on a mostly linear and extractives economic base, and will be a great challenge for scholars, operators, and managers, but will mean a paradigm shift for the science of finance and the law.

Conclusion Remarks

The Selective Incentives are a potential power instrument to develop circular economy and green markets since an adequate tax system. The last concept would only be possible under postulates established in this paper. They are not an exclusive plexus, but they are a guideline to adapt the traditional tax system to circular economy and green markets with a minimum of the traumatic economic situations. Putting emphasis in the behavioral change is the safest way to the triumph of sustainable productive matrix, such as is supported by circular economy and green markets. "Consistent treatment" is a safe theoretical path towards this addressing. On the other hand, the practical implementation of those theoretical categories will depend on economic possibilities of society. Primary structuring elements represent a scientific contribution in order to understand the internal requirement of tax system to influence a behavioral change and, of course, of productive matrices. They are built not as fiscal sacrifice because the effect on the development of a circular productive matrix corresponds to a kind of compensation for indirect amounts that in the fiscal exercise of early stages must suppose the reduction or weighting of the tax rates, as we explained before. The present results are significant in at least two major aspects. First, it allows the design of new tax systems with different meta-goals and the ability to encourage environmentally sustainable productive matrixes. Second, the use of SI theorization helps to understand the structure of tax incentives and their relationship with behavioral changes. Further research should be done to investigate their powerful scopes.

It can therefore be assumed that it will not be an easy task. However, at this point, even the solution will not be. We just have described a sort of theoretical path with many elements, such as issues that need to be considered under the idea to provide more advantages for green markets and circular economy inside economics.

References

Almunia, M. and Guceri, I. (2020). More giving or more givers? The effects of tax incentives on charitable donations in the UK. Journal of Public Economics 183: 104114.

Allayarov, S.A. (2020). Combination of fiscal and stimulating functions of the tax system to ensure financial and economic security. American Journal of Economics and Business Management 3(1): 64–69.

Baker, A. and Murphy, R. (2019). The political economy of 'tax spillover': A new multilateral framework. Global Policy 10(2): 178–192.

Barrett, J.M. and Makale, K. (2019). The environment is not an externality: The circular economy and New Zealand's Tax Working Group. Forthcoming in Journal of Australian Taxation (Special New Zealand Edition).

Bazza, A. (2012). Los beneficios tributarios: Su legitimidad y constitucionalidad como herramienta de políticas públicas. Documentos y aportes en administración pública y gestión estatal I(19): 141–143.

Bengoechea, M.G. (2014). Algunas notas sobre la extrafiscalidad y su desarrollo en el derecho tributario. Revista técnica tributaria (107): 147–170.

Boadway, R. (2019). Rationalizing the Canadian income tax system. Canadian Tax Journal/Revue Fiscale Canadienne 67(3): 643–666.

Braithwaite, V. and Reinhart, M. (2019). Preliminary findings and codebook for the How Fair, How Effective Survey: The collection and use of taxation in Australia, Centre for Tax System Integrity (CTSI), Research School of Social Sciences.

Brimsøe, T.S. and Østerhus, E.M. (2019). Exploring Waste Management in the Circular Economy Concept Trough a Literature Review and a Case Study, University of Stavanger, Norway.

Burman, L.E. and Slemrod, J. (2020). Taxes in America: What Everyone Needs to Know, Oxford University Press.

Cairncross, A. (1978). Keynes and the planned economy. Keynes and Laissez-Faire, Springer: 36–58.

Casado Ollero, G. (1985). Extrafiscalidad e incentivos fiscales a la inversión en la CEE. Hacienda pública española (96).

Cavaleiro de Ferreira, A. and Fuso-Nerini, F. (2019). A framework for implementing and tracking circular economy in cities: The case of Porto. Sustainability 11(6): 1813.

Choi, T.-M. and Taleizadeh, A.A. (2020). Game Theory Applications in Production Research in the Sharing and Circular Economy Era. Taylor & Francis.

Chow, W. and Wang, J. (2017). Capital Gains Tax with Hong Kong Characteristics: Necessity, Feasibility and Design. 6th Annual International Conference on Law, Regulations and Public Policy (LRPP 2017), Global Science & Technology Forum (GSTF).

Culquicondor, R.E.R. (2019). LA IMPOSICIÓN SELECTIVA AL CONSUMO EN EL PERÚ: ENTRE EL FIN RECAUDATORIO Y LA EXTRAFISCALIDAD. Quipukamayoc 27(54): 29–36.

Du, J. and Deng, K. (2017). Getting food prices right: the state versus the market in reforming China, 1979–2006. European Review of Economic History 21(3): 302–325.

Eichfelder, S. and Jacob, M. (2020). Do tax incentives reduce investment quality? Available at SSRN 3515852.

Espadafor, C.M.L. (2011). La determinación de un límite cuantitativo a la imposición indirecta y la extrafiscalidad. Nueva fiscalidad (1): 9–74.

Fontana, G. and Realfonzo, R. (2020). Monetary economics after the global financial crisis: what has happened to the endogenous money theory? European Journal of Economics and Economic Policies: Intervention 1(aop): 1–17.

Ghisellini, P. and Ulgiati, S. (2020). Circular economy transition in Italy. Achievements, perspectives and constraints. Journal of Cleaner Production 243: 118360.

Giraldo-Bedoya, H.F. and García-Duque, C.E. (2019). LA UTILIDAD DEL RACIONALISMO CRÍTICO EN EL CAMPO TEÓRICO Y PRÁCTICO DE LA EDUCACIÓN. 15(I): 91–110.

GOEL, M. (2018). Management of sustainability in fashion supply chain.

Gordillo Pérez, S.M. (2017). Análisis del impuesto a la salida de divisas como tributo extrafiscal: Los modos de extinguir obligaciones en aplicación al impuesto a la salida de divisas en el Ecuador. PUCE.

Graham, J.R. and Smith, C.W. (1999). Tax incentives to hedge. The Journal of Finance 54(6): 2241–2262.

Graham, J.R. and Rogers, D.A. (2002). Do firms hedge in response to tax incentives? The Journal of Finance 57(2): 815–839.

Griffith, R. and O'Connell, M. (2019). Tax design in the alcohol market. Journal of Public Economics: 20–35.

Jia, J. and Li, J. (2019). Brief discussion on the changes of tax system since China's reform and opening up (1978–2018). American Journal of Management Science and Engineering: 26–31.

Justicia, T.E.d. (1993). Hans Werner contra Finanzamt Aachen-Innenstadt. Luxemburgo, Tribunal Europeo de Justicia.

Klemm, A. and Van Parys, S. (2012). Empirical evidence on the effects of tax incentives. International Tax and Public Finance 19(3): 393–423.

Kuah, A.T. and Wang, P. (2020). Circular economy and consumer acceptance: An exploratory study in East and Southeast Asia. Journal of Cleaner Production 247: 119097.

Kyari, A.K. (2019). A theoretical and empirical investigation into the design and implementation of an appropriate tax regime: an evaluation of Nigeria's petroleum taxation arrangements. Tesis Doctoral.

La Rosa, S. (2001). Los beneficios tributarios. Tratado de Derecho Tributario. Roma.

Limonta Montero, R. (2019). La decisión financiera pública y el dimensionamiento del derecho de participación en ella. Apuntes para un debate. Derechos y Libertades 1(41): 323–347.

Lombard, M. and Koekemoer, A. (2020). Conceptual framework for the evaluation of sugar tax systems. South African Journal of Accounting Research: 50–62.

Makarova, I. and Shubenkova, K. (2020). Features of Logistic Terminal Complexes Functioning in the Transition to the Circular Economy and Digitalization. Modelling of the Interaction of the Different Vehicles and Various Transport Modes. Springer: 415–527.

Martín Queralt, J. and Lozano Serrano, C. (2015). Curso de Derecho Financiero. Madrid, España, Tecnos.

Musgrave, R. and Peacock, A. (1967). Classics in the Theory of Public Finance, St.

Musgrave, R.A. (1969). Teoría de la hacienda pública.

Musgrave, Richard and Peggy b. Musgrave. (1980). Public Finance in Theory and Practice, 1980.

Musgrave, R. and Musgrave, P. (1992). Hacienda Pública: Teoría Y Práctica. Edited by José María Lozano Irueste.

Neumark, D. and Wascher, W. (1994). Employment effects of minimum and subminimum wages: Reply to Card, Katz, and Krueger. ILR Review 47(3): 497–512.

Night, S. and Bananuka, J. (2019). The mediating role of adoption of an electronic tax system in the relationship between attitude towards electronic tax system and tax compliance. Journal of Economics, Finance and Administrative Science.

Olson, M. (1965). The Logic of Collective Action. Public Good and the Theory of groups. Cambridge, Massachusetts, United States of America, Harvard University Press.

Parchomenko, A. and Nelen, D. (2019). Measuring the circular economy—A multiple correspondence analysis of 63 metrics. Journal of Cleaner Production 210: 200–216.

Rezzoagli, C.L. (2006). Beneficios tributarios y derechos adquiridos, Cardenas Velazco.

Roca, J. (2010). Evaluación de la efectividad y eficiencia de los beneficios tributarios. Inter-American Bank Review.

Santagata, R. and Zucaro, A. (2020). Assessing the sustainability of urban eco-systems through Energy-based circular economy indicators. Ecological Indicators 109: 105859.

Sarin, N., Summers, L. et al. (2020). Tax reform for progressivity: A pragmatic approach. Tackling the Tax Code: Efficient and Equitable Ways to Raise Revenue.

Sicuelo, I.R. (2012). La exención en el impuesto al valor agregado de la educación privada de idiomas, Universidad Nacional de Mar del Plata; Consejo Profesional de Ciencias.

Tomasetta, C. and Zucchella, A. (2017). The Life Cycle Sustainability Assessment Approach Applied to Tangible Cultural Heritage Conservation—Developing a Support Instrument for Cultural Heritage Management within a Circular Economy and Life Cycle Thinking.

Tørsløv, T.R. and Wier, L.S. (2020). Externalities in International Tax Enforcement: Theory and Evidence, National Bureau of Economic Research.

Ugalde, A.F. (2019). Influencia romano-germánica en el derecho tributario latinoamericano y, en particular, en la doctrina chilena. Análisis sincrónico y diacrónico para la discusión dogmática de hoy. Vniversitas 68(139).

van Arendonk, H. and Engelen, F. (2005). ARENDONK, Henk van, et al. Hughes Lasteyrie du Saillant: crossing borders? A Tax Globalist, Essays in honour of Maarten J. Ellis. H. van Arendonk and F. Engelen. Groningen: 181–209.

Varona Alabern, J.E. (2009). Extrafiscalidad y dogmática tributaria. Madrid, Marcial Pons.

Varona Alabern, J.E. (2010). Los tributos extrafiscales. Extrafiscalidad regular e irregular. Tratado sobre la Ley General Tributaria. Homenaje a Álvaro Rodríguez Bereijo 1.

Velarde Aramayo, M.S. and González García, E.P. (1997). Beneficios y minoraciones en derecho tributario. Madrid, Marcial Pons.

Vona, F. and Marin, G. (2019). Measures, drivers and effects of green employment: evidence from US local labor markets, 2006–2014. Journal of Economic Geography 19(5): 1021–1048.

Yacolca, D. (2019). LA EVOLUCIÓN DEL CONCEPTO DE TRIBUTO HACIA FINES EXTRAFISCALES. Revista Argumentum-Argumentum Journal of Law 14: 225–230.

Yan, L. (2017). Simulation study on tax system design of real estate tax. Construction Economy (6): 14.

Zhu, M., Li, K.X., Shi, W. and Lam, J.S.L. (2017). Incentive policy for reduction of emission from ships: A case study of China. Marine Policy 86: 253–258.

CHAPTER 6

The Importance of Business Associations in Promoting the Socio-Environmental Sustainability Innovation Concept

Araceli Regalado-Cerda

Chapter Objective

The chapter explores the importance of business associations in promoting the socio-environmental sustainability innovation concept among their members so as to include it in their business models to prevent, mitigate, and eliminate probable damages to the environment and society caused by their business activities. The chapter begins with some considerations to the innovation, sustainability, and associativity[1] concepts. Then, it presents the results of empirical research among national franchise associations during 2018 and 2019 to evaluate their interest in promoting the sustainability concept among their members for their awareness and to increase their commitment towards the environment and their community.

Introduction

In recent studies and despite significant progress, the incorporation of environmental and social sustainability initiatives in the business sector is restricted to the compliance of legal requirements and responsibilities, as this incorporation is associated with the idea that it involves expenses and costs with no immediate or short-term benefits, as well as reducing profitability for the businesses involved. Similarly, many businesses

Istanbul Aydin University, Amores 1625 – 102B, Col. Del Valle, Alcaldía Benito Juárez, 03100 Ciudad de Mexico, Mexico; aregalado37@hotmail.com

[1] Associativity: strategy of collaboration, which is linked to specific businesses; it is a tool at the service of businesses. Companies, in this context, develop a collective effort for the achievement of common objectives, which can be very dissimilar, like buying research and development programs or better positioning in the value chain for your business trading scheme (Poliak 2001).

consider this socio-environmental dialog as a "topic du jour", as a fashion trend to improve firms' image, but with no real commitment to protect the environment, and, even less, to fulfill social and community responsibilities. Nevertheless, there is some evidence that well-established and well-positioned companies, as well as others of recent creation, have incorporated innovative sustainable strategies in their business models, e.g., products, services, organization, and processes to meet new market demands and government regulations (Haanaes et al. 2011), as well as, social and environmental issues endorsed by different organisms, at local and international levels.

Today businesses face the commitment to establish ethical and responsible business models, conscious of the probable negative impact that their activities could damage the environment and their communities, considering the framework of international treaties signed by most nations, like the World Pact of 1999 (Corporate Social Responsibility), the 17 United Nations (UN) Sustainable Goals of 2015 (Sustainable Development), and the Paris Agreement of 2015 (Climate Change), among others.

Considering that sustainability and innovation are highly related in the business sector, as innovation has become an essential element to achieve sustainability, and at the same time, firms are incorporating a sustainable vision into their innovation process, we can conclude that sustainability is a vital element for innovation as well as the other way around (Layrisse and Madero 2018). So, the sustainable innovation concept, recently created, defines those innovations that eliminate, reduce, and mitigate the negative impacts on the environment and the society produced by business activities, including their stakeholders (employees, customers, consumers, suppliers, competitors, investors, lenders, insurers, non-governmental organizations, media, governments, and society overall). In other words, they are innovations that create positive and significant impacts on the reduction of socio-environmental negative impacts due to the form in which businesses create, deliver, and capture value (Bocken et al. 2014).

Massachusetts Institute of Technology Sloan School of Management (MIT Sloan) and the Boston Consulting Group (BCG) have joined forces since 2009 to learn the business implications of sustainability, the corporate commitments to sustainability-driven management, and to increase knowledge about business adoption of sustainable practices, and to support the integration of sustainability into business strategy (Berns et al. 2009, Haanaes et al. 2011, Kiron et al. 2017). Thus, it is of the utmost importance to explore which and how the incorporation of socio-environmental sustainable innovations have been in different industries, and in their business models, e.g., products, services, processes, and marketing, as well as which the challenges, risks, and the benefits of this incorporation have been to sustain their competitiveness, growth, and development in order to intensify the incorporation of this concept in the business world. A possible means for spreading the knowledge, commitment, and benefits of the sustainable innovation concept could be through business associations that represent an industry or sector, so as to profit from the opportunities of knowledge sharing that this associative model offers. Consequently, the following hypothesis was considered for this chapter: if business associations promote the socio-environmentally sustainable innovation concept among their business members, then their members would consider the incorporation of the concept in their business models to prevent, mitigate, or eliminate socio-environmental damages caused by their business activity.

In order to develop an exploratory research on the subject, the franchise sector was chosen to exemplify a business sector that could strengthen its sustainability knowledge and commitment through its most representative business association, in this case, the national franchise association.

Conceptual Framework

Innovation from a Business Perspective

Business literature offers a wide range of definitions around the construct of innovation, without being able to fully delimit it, given the complexity of the innovative process in business (King and Anderson 2003 cited by EOI 2007). But here is a clear one declared by Joseph A. Schumpeter "the introduction of a new good (product) for consumers or of higher quality than the previous ones, the introduction of new production methods for an industry sector, the opening of new markets, the use of new sources of provisioning, or the introduction of new ways for competing that lead to a redefinition of the industry (sic)", differentiating from a mere creation without considering the business world and innovation in business (Schumpeter cited by Montoya 2004). Later in 1942, Schumpeter conceived the term Creative Destruction to describe the dismantling of long-standing practices in order to make way for innovations in the manufacturing process to increase productivity (Schumpeter cited by Kopp 2019).

Additional remarks, associated with innovation in business, are: Mintzberg et al. (1997) added the technological element describing an innovative organization as one that faces complex technologies or systems constantly; Porter (1998), the marketing guru, ties innovation with competitiveness in the search for a favorable position within an industry; later, Porter conceived the term Shared Value, which promotes business policies and practices to improve economic and social conditions on their community to add a competitive advantage; Leff (2008) introduces the environmental element to innovation in business by stating that the reconstruction of the organizational forms and processes based on environmental rationality is possible, and it implies rethinking a new economy that is based on ecological potentials and technological innovation; for Albornoz (2009), there is a need to "examine the evolution of innovation and review the experience in terms of indicators that evaluate their effectiveness", that is to apply measures (key productivity indicators—KPI) that could evaluate the usefulness and effectiveness of the innovations involved; Thompson et al. (2012) indicate, that thanks to technology innovations, organizations can address the opportunity to start new businesses and the society to access new industries. KPMG (2017) summarizes that innovation is a strategic weapon and a philosophy that permeates throughout the organization, and it must be promoted as an absolutely necessary process, "on which the survival or the disappearance of the business depends".

The Organization for Economic Cooperation and Development (OECD), together with other organizations, positions innovation as "the main driver of socio-economic growth in the long term", associating it with the creation and dissemination of products, processes, technology, and non-technological methods. Likewise, OECD considers innovation as a key element in the search for viable alternative solutions to the main sustainability problems in the form of social, economic, and environmental difficulties that all nations face in developed or emerging environments (OCDE 2012, KPMG 2017, EOI 2007). Correspondingly, the 2017 MIT Sloan and BCG research, among "thousands of managers and more than 150 executives and thought leaders interviewed", proved that sustainability can be a driver of innovation, efficiency, and business value, and mentions that this interest fluctuates according to geographies and industries (Kiron et al. 2017). See Table 6.1 describing the different types of innovation included and defined in the Oslo Manual of 2018.

Table 6.1. Types of innovation according to the Oslo Manual.

Types of innovation according to the Oslo Manual	
Type	**Description**
Product	Introduction of a new product or a significantly improved product, or service, considering its characteristics, previous use, significant improvements in techniques, materials, software, ease of use.
Process	A new production or availability method or significant improvements, including new techniques, equipment, or software.
Marketing	New methods in marketing, including significant changes in product design, packaging, merchandising, promotion, and price.
Organizational	A new organizational method in marketing, workplace, and outside relationships.

Source: Own elaboration, based on OECD/Eurostat (2019).

Innovation Applied to Business

The 2017 KPMG (Mexico) research among 864 corporate executive chairmen (CEO) concluded that 75% of respondents considered innovation as one of their three key priorities to improve their competitiveness, and 51% of respondents consider innovation as a necessary strategy to increase competitiveness. Thus, innovation and "the improvement of the innovation capacity" have been placed among the most important variables considered by business executives, e.g., cost reduction, customer satisfaction, development of efficient systems, as the factors that will impact their profitability in the near future. Managers consider innovation as the backbone of their priorities, challenges, and organizational development actions in the short term, and most probably, innovation will be the key element for their long-term success and competitiveness (KPMG 2017).

Companies employ, basically, three types of innovation: *incremental innovation* (improves performance through small and steady changes), *radical innovation* (focuses on doing things differently), and *disrupt innovation* (creates a new market and value network with simple, accessible, and affordable solutions) (Searcy 2018, Christensen 2020).

Innovation in business models is of the utmost importance to maintain a sustained strong market position, and its research is vital. Business model innovation is defined as the search for improved products, processes, and how products are delivered to customers or to new markets to obtain new revenues and competitive advantages. Advocates for this type of innovation state that innovation in business models generates a higher correlation with the growth in the operating margin than only innovation in the product development or process reengineering (IBM 2006, Márquez 2010). In a qualitative multi-case comparison survey to explore business model innovations, Barjak et al. (2013) resume innovation in business models as:

1. New value propositions that, in most cases, coincide with product innovations.
2. Radical product innovations that coincide with new customer value propositions.
3. Changes of business systems, processes, internal and external organization, and division of labor along the value chain.
4. Radical process and organizational innovations that coincide with new business systems.
5. A new approach for capturing value that coincides with a process/marketing innovation.
6. Radical process and marketing innovations will more often coincide with new revenue models than incremental innovations.

Additional consideration of innovation in business is the systemic approach for innovation, defined as a "set of elements and relationships which interact in the production, diffusion, and use of new and economically valuable knowledge (…) and are either located within or rooted inside the borders of a nation-state" (Lundvall 1992). Peraza and Mendizábal (2016) also considered innovation from the systemic approach, and studied Freeman who defines it as "the network of institutions in the public and private sectors whose activities and interactions initiate, import, and disseminate new technologies" (Freeman 1987 cited by Peraza and Mendizábal 2016).

At this point, we can also refer to the attractive concept of the business innovation ecosystem, which refers to stakeholders (customers, suppliers, staff, specialists, advisors, mentors, employees, competitors, investors, lenders, insurers, non-governmental organizations, media, governments, other firms in or outside the sector, as well as the society overall) linked to a company, interacting with each other through vital processes, to explore and introduce new ideas in the business model to add value and obtain competitive advantages that allow to consolidate its market position or solve any other issues that might affect the group of stakeholders.

There is a growing interest in ecosystems, as there is a necessity to acquire and coordinate diverse and novel capabilities; at the same time, organizations must review the myths and realities of business ecosystems before committing themselves to an ecosystem's strategy. Fuller et al. (2009) elaborate on 10 myths and realities of business ecosystems to avoid misunderstandings when participating in one. A healthy innovation ecosystem will require favorable conditions for innovative processes, fluid communication channels, updated management information regarding market trends, personal competition and, above all, solid leadership to motivate stakeholders to participate in the development of innovative ideas and processes within and outside the network. We can refer to this interesting concept with two approaches: first, as a set of interactions among a diversity of actors leading towards good things (energy, money, and ideas productively interacting with one another); and second, as a set of partners aligned for the desired value proposition to materialize, which requires strategy, leadership, collaboration, communication, accountability, and ownership (Adner 2016). Blanda (2019) believes we live in the era of the ecosystem, "as a series of interconnected, bi-directional relationships by multiple partners". Sean Blanda elaborates that each node of the ecosystem requires to be worked, documented, valued, and prioritized. Accordingly, a new business figure has been created, the Partner Manager, who needs to not only be good at creating partnerships, but also building and connecting all of the infrastructure required to scale a partnership program, an "EcosystemOps". Figure 6.1 is an example of the innovation ecosystem in the franchise sector.

Additionally, it is possible to classify innovation according to the eco-business models′ concept defined by Bocken et al. (2014), which includes the sustainability element, not previously considered in business models. With this in mind, eco-innovation refers to "those innovations that generate improvements in the management of the environmental impacts of production and consumption activities, and have been shown as a key piece to mitigate the traditional dichotomy between competitiveness and sustainability" (Carrillo-Hermosilla et al. 2016). Table 6.2 presents the Bocken et al. eco-innovative business models.

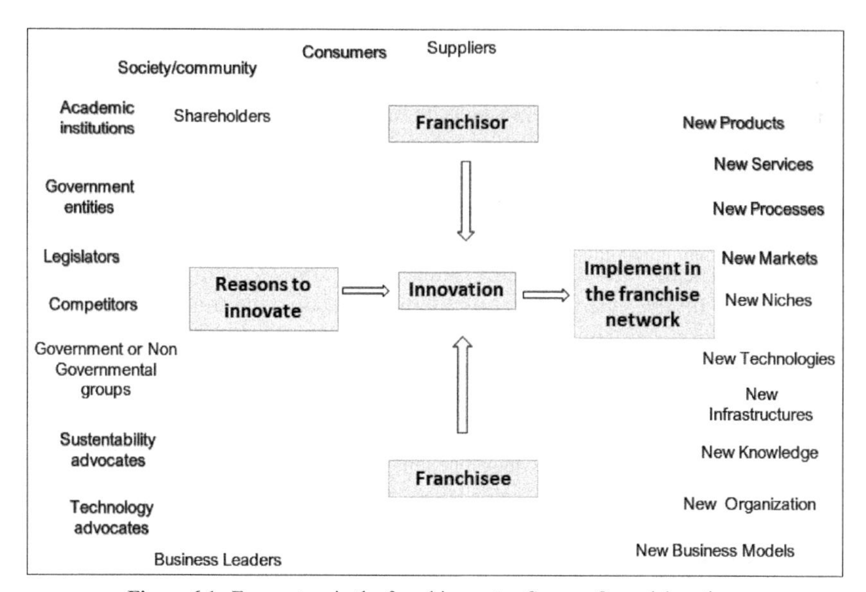

Figure 6.1. Eco-system in the franchise sector. Source: Own elaboration.

Table 6.2. Eco-innovation business models.

Eco-innovation	Description
Technological	Maximize material and energy efficiency.
	Development of waste value.
	Substitution of traditional processes for sustainable and renewable processes.
Social	Ownership (products and services efficiency).
	Responsible consumption (conscious and sufficient).
Organization	Reconsideration of the business responsibility before society and the environment.
	Development of improved solutions.

Source: Own elaboration based on Bocken et al. (2014).

Much more can be added and studied on the ample concept of innovation in business and in business models, among others, functions, motivators, processes, leadership, training, indicators, etc., but let us focus on other considerations for the chapter objective.

Sustainability

It is important to understand the sustainability concept, so here are some thoughts to clarify the construct, starting with Adams (2006), who traces back to 1969 the origin of the idea of sustainability, concerning the "perpetuation and improvement of the living world—the natural environment of man, its resources—believe in all living beings", alluding to the management of "air, water, soils, minerals, and living species, including man, to achieve the highest possible quality of sustainable living" (IUCN 1969 cited by Adams 2006). This concept was extensively discussed at the United Nations Conference on the Human Environment celebrated in Stockholm 1972, the World Conservation Strategy at Geneva 1980, The Brundtland Commission on 1987, the United Nations Conference on Environment and Development at Rio de Janeiro 1992, as well as by national governments, trade leaders, and different non-governmental organisms, to become a major worldwide concern.

The document, Our Common Future (UN 1987), better known as The Brundtland Report, was the first serious attempt to resolve the confrontation between development and sustainability. The report provides the first definition of sustainable development, noting that "it is in the hands of humanity to ensure that development is sustainable, e.g., to ensure that it meets the needs of the present without compromising the capacity of future generations to satisfy their own".

Since then, the sustainable development concept has emerged as the guiding principle of long-term global development, with the aim of achieving a balanced approach to economic-social development and environmental protection, e.g., the three pillars or basic dimensions encompassed by the concept has been graphically expressed in different ways according to different perspectives as well; of these, the most well-known and used is the proposal by the IUCN Program for 2005–08, based on a model of interlaced circles that sought to express the need for integration of the three pillars to restore the balance between sustainability dimensions (Figure 6.2).

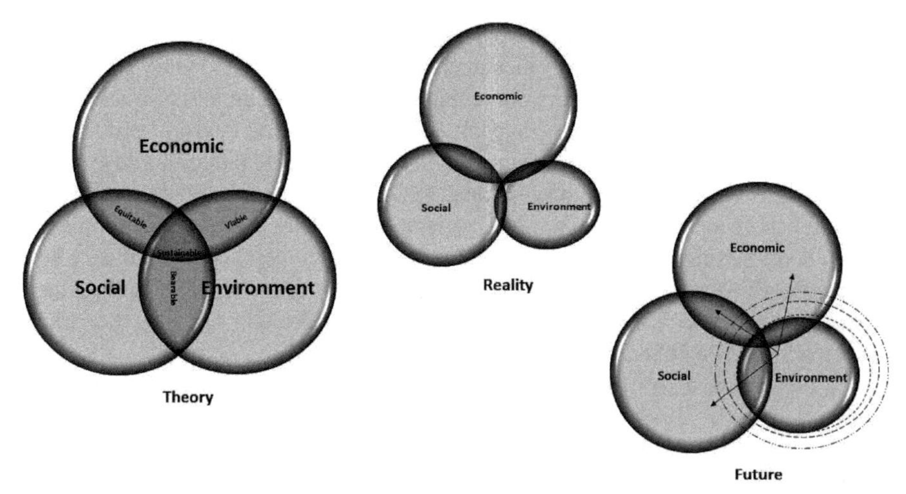

Figure 6.2. Sustainable development according to UICN. Source: Own elaboration, based on Adams (2006).

The reading of Figure 6.2 displays that despite the efforts that have been achieved since 1969 ***Theory***—economic development continues to prevail today over social development and the preservation of the environment—***Reality***; nevertheless, it is expected that the efforts for the environment care in the future match with the theory—***Future***.

At the United Nations Summit on Sustainable Development celebrated in New York on 15 September 2015, 193 UN state members signed the document entitled Transforming Our World: the 2030 Agenda for Sustainable Development, which included 17 UN Sustainable Development Goals, unraveled in 269 goals, to tackle poverty, combat inequality, reduce injustice, and mitigate climate change, in favor of the people, planet, and world prosperity (UN 2015).

The sustainable development concept has been addressed from different approaches, including changing values, moral development, social reorganization, transformational processes, business ethics, business strategy, investment opportunities, human rights, and, in all cases, from the perspective of a promising vision of the future.

Over time, sustainable development has ceased to be a novel concept and is being applied in all areas of human work, including business for achieving the category of a strategic business approach to environmental protection, and social improvement in terms of equity and community support, becoming an essential element for any company wishing to maintain or improve its market position.

Sustainable development refers to the process to attain sustainability, so sustainability is the goal to be achieved.

Sustainability in Business

Freeman's stakeholder theory addressed the importance of why companies should work towards sustainability, as it is in their best economic interest to strengthen their relationship with stakeholders to meet their business objectives. Pojasek agrees with Freeman in the importance of stakeholders for businesses while protecting, sustaining, and enhancing the environmental, social, and economic resources needed for the future (Freeman 1984 and Pojasek 2007 mentioned by Layrisse and Madero 2018). Then, sustainability is for business organizations the goal that should translate into "the awareness that their professional function is to create wealth and quality of life for the people who constitute it, generating the least possible environmental impact in the setting in which it operates", that is, in alignment to the business effort to achieve profits and well-being. Then, sustainability refers to new management processes that define "the values and principles, as the role that its organization plays in society" (Blasco 2008 cited by Vilaseca 2008).

In this sense, "the sustainable enterprise concept has been shaped as one that creates economic, environmental, and social value in the short and long-term contribution to the well-being and the true progress of present and future generations" (Rodríguez 2012).

The Sustainable Development in Mexico 2018 research developed by KPMG (Mexico), among 143 top managers, concluded that sustainability should be understood as a strategy of cross-cutting concerns for companies affecting society and the community; concerns that are part of a whole where natural resources, international trade, infrastructure, and technological advances converge, as well as others that constitute the future of the economy (KPMG 2018).

In order to move towards being a sustainable development entity, the concept of Corporate Social Responsibility (CSR) emerges for businesses. CSR establishes commitments to minimize the negative impacts of their operating activities, based on open and constant communication with their stakeholders. The best examples of CSR have created better scenarios for organizations, society, and the environment, as their sensitivity to ecological and social challenges has a beneficial impact on their development (Navarro 2012). For Porter and Van der Linde (1995), business organizations begin to view CSR best practices as a competitive advantage that facilitates their penetration into markets. This is confirmed by several researches completed by MIT Sloan and BCG, KPMG, and Deloitte mentioned in this chapter.

The practice of the CSR requires moral conduct adhered to the mandatory norms of law and ethical conduct for its fulfillment by own choice, and not by external cohesion like a mandatory government normativity. Accordingly, the CSR is perceived as the firm commitment to help and solve social and environmental problems in terms that exceed what is legally stipulated, which in the short, medium, and long term should add value to both the company and the society itself; the term Shared Value endorses policies and practices that increase the competitiveness of a company while improving the economic and social conditions in its community (Porter and Kramer 2011). Professional experience shows that "the businesses that endure are the ones that establish fair relationships in which all actors win" (KPMG 2018). A socially responsible company is one that assumes citizenship as part of its purposes, bases its vision and social commitment on principles and actions that benefit its business, and positively impacts the communities in which it operates.

In 2020, during the annual World Economic Forum (WEF) annual meeting held in Davos, Switzerland, a number of disagreements erupted in the business, economic, financial, and political communities in regards to the efforts being developed to hold the damages to the environment, and consequently to vulnerable societies, less developed countries, and, in general, to the world. Two voices spoke loudly and represented the opposite positions, one from the young Greta Thunberg denouncing "Businesses and governments' leaders offer empty words and promises rather than action. Unlike you, my generation will not give up without a fight, we are telling you to act as if you loved your children above all else." And the other voice was from Donald Trump, who attacked environmental alarmists and economic pessimists, arguing that the "perennial prophets of doom and their predictions of the apocalypse" should be rejected (Edgecliffee-Johnson and Sevastopulo 2020).

At the same WEF annual meeting, other voices, from bank executives of Citibank, Goldman Sachs, and others, rejected the suggestions that they should refuse to work for clients that were significant polluters. They argued that banks are not to draw the line or tell the companies or enforce standards on big polluters. On the other side, there is a growing number of investors dissatisfied with the "profit-maximization at any cost" view, and focus on new long-term accountability on the environment and society. This view was represented by executives from AIG, UBS, BlackRock, Avaaz, Japan's Government Pension Investment Fund, among others. They agree that it is imperative for asset managers to support green finance to contribute to society, even irrespective of consumer pressure or government rules, and at the risk of losing their company support (Tomlinson 2018, Fink 2020).

At the same time, evidence shows the negative effect of environmental events in funds and stocks, e.g., rising sea levels pose long term risk to the sovereign credit ratings of countries with large areas at risk of submersion. The economic and social repercussions of lost income, damage to assets, health issues, and forced migration from unexpected events are immediately felt in farming, tourism, and trade mainly in emerging economies, but also in developed economies, even though the latter hold economic mechanisms to restore the damages caused by environmental events (Daily News 2020). During late February 2020, another example of financial crisis showed how stock markets suffered the coronavirus jitters as it sparked sell-offs because of an uneasy sentiment affecting the travel, tourism, and hospitality sectors in Europe and Asia, as well as other sectors affected by the CV19 virus (Lockett and Geordiadis 2020).

The above mentioned are examples of a cultural crisis. Safran (2019) sees this as a crisis of belief, an impossibility of cooperation across countries and within countries, as the (economic) interests motivated against solving the problem are more powerful than the interests in solving it. Then, Safran requests for solidarity initiatives through small collective sacrifices, because there is a chance that it might already be too late to avoid runaway climate change.

Additional voices alert that the present and future focus will be on environmental, social, and governance issues (ESG), as well as on sustainability. Consequently, business ethics will pledge to serve multiple stakeholders with different interpretations (shareholders, workers, customers, suppliers, communities, among others), and it can mean taking tough decisions on investment opportunities in pursuing their purpose (Crow and Tett 2020). Since 2009, a global survey of more than 1,500 corporate executives declared a strong consensus that sustainability was having—and would continue to have—a material impact on how companies think and act, in terms of investments, as these are to comply with government legislation, consumer concerns, and employee interests (Berns et al. 2009). As of 2020, sustainability continues to be the tone, and will continue to be for the next years, dividing the world's wealthiest investors with consequences to the remaining world in the environmental and social issues.

Thus, corporate sustainable practices are essential for a strong market environment and an enduring society (Searcy 2018). In order to accomplish business sustainable development goals, it is necessary to review internal business activities to develop new capabilities and characteristics, develop a new culture, process and efficiency measurements, financial models, and skills to search for internal and external interaction, and collaboration within their sector (Berns et al. 2009). Consequently, cooperation plays a major role in the implementation of sustainable development principles, activities, and goals within the business organizations (Çörtoğlu 2019).

Business Associative Models

As stated previously, the present and future focus will be on environmental, social, and governance issues (ESG) and sustainability. Accordingly, sustainability has turned into a key element to business strategy and management, and the risks of failing to act decisively and on time are growing. Hence, business leaders will have to change their strategies to meet the new business environment.

In an interesting analysis of philosophy and strategy, Roy and Purkayastha state the need to continuously change business strategies:

"Heraclitus, considered to be the most prolific thinker among the philosophers of Pre-Socrates era, had said that the universe is in a flux and "you can't step in the same river twice". This to strategy means that the same strategy will not work for a firm infinitely. We cannot take anything for granted and, from time to time, it is necessary to change strategy (sic)" (Roy and Purkayastha 2007).

Globalization, innovation, sustainability, accelerated change processes in highly competitive environments dominated by consumers with increasing demands, and the new government regulations imposed on the business sector, are characteristics of the present and the future scenario, where business organizations must implement strategies and mechanisms to increase their competitiveness, productivity, and internationalization (Mendoza 2011). Here, any kind of business integration and collaboration, such as business associations, business networks, business ecosystems, and similar groups can aid companies in acquiring knowledge to develop and establish strategies to face this business environment to meet new markets and government demands for their own survival and expansion.

Business associations are considered to be a means or a tool for stimulating the business systems that wish to sustain and increase their market position through the development of agreements, strategic alliances, the articulation between businesses of different sizes, productive relationships, and networks (Liendo and Martinez

2001). In other words, business associations are collaborative strategies in terms of shared spaces and knowledge of a particular type of business (Poliak 2001, Alarcón 2015). In this way, business associative models are perceived as a practical tool for business development, growth, and sustainability, as they improve technological, technical, economic, innovative, productive, and training conditions, which impact on the levels of employability, productivity, and economic growth of the country (Alarcón 2015).

Additionally, business associations design ethical behavioral guidelines for their memberships, and the sector they represent. Generally, these business associations are guided by an elected leader, the executive team, and the council (board of directors). The responsibility of the leadership team is to maintain and work on behalf of its members' vision, mission, and drive, by creating strategies according to actual economic, legal, social, and government circumstances affecting their sector. At the same time, this team needs to be in permanent communication with its members, and collaborate with other groups, such as government and non-government organisms, academics, and even other business associations, to develop valuable information and knowledge on behalf of their members and their sector's benefit.

As mentioned, business associations design and supervise ethical guidelines agreed by their members and in favor of their own community. Here, sustainability fits perfectly within this commitment, as it is closely linked to business ethics on how businesses create, produce, perform their activities, and deliver value. Consequently, business associations should accept their own accountability to protect and take care of the environment, on community health and safety issues, ethical business performance of their members, and, in general, the improvement of quality of life of their local communities, and avoid damages to the environment.

Accordingly, the top management of business associations need to design sustainability approaches to endorse, promote, evaluate, and recognize efforts from their memberships. To achieve this, it is recommended that their executive team and council develop and implement a sustainability business case strategy: (1) assign a leader responsible for this effort; (2) develop socio-environmental informative and training events for their memberships; (3) design return on investment methodologies and goals; (4) promote investments in sustainable projects to reinforce and update business models; (5) display socio-environmental, financial, and non-financial benefits; (6) establish key indicators to monitor progress; (7) provide required coaching for implementation; (8) recognize sustainable efforts and accomplishments from their members to promote and strengthen sustainability commitment.

Considering associations, a form of networks, Regalado-Cerda (2019a) explains that networks are social spheres offering the possibility to establish solid and strong relationships with high levels of engagement, and work for the accomplishment of their own objectives. This success depends upon the continuous knowledge sharing and problem-solving among their members. For this, business associations should recognize and take advantage of the resources, capabilities, and abilities of each of their members to try to complement each other through a collective and accumulative knowledge sharing process. This associative resource offers the possibility of creating and implementing innovative solutions in favor of the common welfare of its members.

This is especially possible when the network is complex and heterogeneous, because as feedback increases, it generates ideas and solutions from the members' previous and shared experiences.

Considering the characteristics of the network, the relationships among their members, and their performance in reaching goals, Rost (2011) introduces three types of networks, which can be applied to business associations as well:

1. **Close and Stable.** They are in search of relationships, processes, and a solid exchange of experiences.
2. **Open and Professional.** They generate and share the knowledge that stimulates innovation and problem-solving within the network.
3. **Complex and Active.** They recognize and access the resources and capabilities of their members.

It is important to highlight that the stronger the network process of integration, collaboration, and knowledge sharing, the stronger and robust will be the process to reach for innovative solutions for the well-being of its members.

A business ecosystem can also be considered as an associative model, as they are models where knowledge develops and is shared. We can work with the idea that a business ecosystem is the community where businesses and their stakeholders know their environment, interact through different processes, present, explore, and introduce new ideas to the business model to increase value and to obtain competitive advantages to consolidate their market position and differentiate from their competition.

Even though associativity models are increasingly frequent in all types of organizations, particularly, in small and medium-sized enterprises, it is still an emerging phenomenon scarcely analyzed (Rosado et al. 2018). This chapter aims to reflect the importance of business associations to increase the consciousness on the sustainable concept among their associates.

Innovation, Sustainability, and Associativity Models

Regarding the previous remarks on the major concepts of this chapter, innovation, sustainability, and associative models, we can recap the following:

First, innovation is a crucial element for the business sector to maintain their competitiveness and develop competitive advantages, and it is also an essential element in the search for solutions to the major social, economic, and environmental problems being faced by all nations.

Second, sustainability is considered the present and future main component to be included in the business models to meet new market demands, consumer expectations, as well as to comply with governmental regulations and standards.

Third, different business associative models can support the business sector to achieve sustainable goals through a strong knowledge sharing process among their members and/or stakeholders. The business executive team collaboration with different stakeholders (internal and external, here associations), together with a clear vision for a sustainable business can create successful business strategies, in this case, successful sustainable strategies in favor of the environment and their community.

It is important to consider that innovation and sustainability are highly related in the business sector, as innovation has become an essential element to achieve sustainability. At the same time, firms are incorporating a sustainable vision into their innovation process, so we can conclude that sustainability is a vital element for innovation as well as the other way around (Layrisse and Madero 2018). Accordingly, a new concept has been created: sustainable innovation, which defines those innovations in business that eliminate, reduce, and mitigate the negative impacts on the environment and the society (stakeholders) produced by business activities. In other words, innovations create positive and significant impacts on the reduction of socio-environmental negative impacts due to the form in which businesses create, deliver, and capture value (Bocken et al. 2014). Businesses that are interested in fulfilling sustainable goals will necessarily have to make innovative changes in their products, servicing, manufacturing processes, logistics, in other words, in their business models (mission, vision, value propositions, market segment, customer relations), and within their organization (leadership, training, personnel capabilities and characteristics, infrastructure, processes, financial investments, activity measurements).

In order to achieve new goals, in the business world, multi-stakeholders' considerations need to be contemplated. Here, the executive team will have to create and develop new strategies for their businesses with the stakeholders' participation. At this point, it is important to highlight the employees' contribution as an important source for companies to obtain vital information and cooperation to become innovators in sustainability; employees will require to be motivated and incentivized to actively support the company's new goals. The executive team will require to define a clear sense of purpose to build a new culture with goals, assign budget and resources, define values, allow a collaborative environment with other stakeholders, provide positive reinforcement, and establish measures of accountability for the social and environmental value creations (Geradts and Bocken 2018)

An important resource for the executive team to obtain vital information of their own sector is the participation in business associations, where plurality and respect flow to create high-performance teams and offer the opportunity to obtain and share market, and government information with other executives of the same sector. Generally, these business associations are non-profit organizations created to establish ethical norms for the associates and the business sector; promote, defend, and represent their sector before the market and the government by offering its associates benefits and services to help them improve their business by raising quality standards, reducing costs, and facing government demands and regulations, among others.

Compiling the chapter's three main concepts, we can state that business associations are an interesting tool for business executives to develop consciousness of the sustainability concept, and to develop strategies to accomplish business sustainable goals. At the same time, sustainability goals will only be achieved through the creation of sustainable innovations (new products, services or processes, and business models) to benefit the environment and the society by preventing, eliminating, and mitigating damages caused by the business activity, even if these innovations are incremental, radical, or disruptive.

This last statement stirred the present work to explore if business associations are interested in the socio-environmental sustainability concept to promote it (create consciousness) among their members for their own advantage (e.g., improve their company image, create competitive advantages, increase profitability, enhance employee loyalty and acceptance, stakeholders support, among others), so as to incorporate the concept in their business models to support the achievement of local environmental and social goals in favor of their businesses and their communities.

Case Study

Introduction

According to the 2017 research performed jointly by the Massachusetts Institute of Technology (MIT) Sloan Management Review (SMR) and the Boston Consulting Group (BCG) among tens of thousands of managers and 150 interviews (executives and thought leaders), in order to learn about the business adoption of sustainable practices and to support the integration of sustainability in business strategy, one of the conclusions was that corporate leaders remain a minority and are unevenly distributed across geographies and industries. At the same time, this survey presented a handful of standout companies that are demonstrating that sustainability can be a driver of innovation, efficiency, and lasting business value, together with 8 key lessons to drive sustainable business practices, regardless of industry or region (Kiron et al. 2017). These lessons will be referred to, along with the remarks of the empirical research for this chapter.

In 2019, Regalado-Cerda presented an exploratory research to determine the level of inclusion of the sustainability concept in 43 Mexican franchise brands. The analysis exhibited that 64% of participants mentioned that they included the concept in their business activities, but when asked for specific activities, they do not include additional economic, organizational, or process efforts; in short, they are isolated efforts that mainly comply with legal requirements to avoid fines/sanctions, or are activities developed to access tax exemptions and investment opportunities, and/or to improve brand image recognition. Another conclusion was that there was no environmental awareness/consciousness in the sector, nor strategic value consideration to reflect on the sustainability factor as a critical factor for their businesses. Needless to say, there was not a clear understanding of the concept. This exploratory research motivated the author to further research on what the national franchise association was doing to promote the concept, and in turn to study what the national franchise associations around the world were doing in favor of the sustainability concept (Regalado-Cerda 2019b).

Survey Main Objective

In order to research the chapter hypothesis, which is: *If business associations promote the socio-environmentally sustainable innovation concept among their business members, then their members would consider the incorporation of the concept in their business models to prevent, mitigate, or eliminate socio-environmental damages caused by their business activity*, an exploratory research was conducted in the franchise sector, as an example of a business sector. This sector has been very successful with economic and social contributions in most countries, where it is well-positioned. Important franchise national business organizations were chosen to analyze their interest in the sustainable innovation concept (previously described), and whether they were exerting sustainability strategies with their members.

Briefing. Two empirical questionnaires were conducted in the franchise sector during July 2018 and July 2019 among the 43 national franchise associations, members of the World Franchise Council, to validate if they had any interest in promoting the socio-environmental sustainability concept among their members so as to encourage its introduction in their business models (Regalado-Cerda 2019a).

Instrument. The questionnaires were sent to the top management of these national associations through the Survey Monkey application. The following items questioned the national franchise associations on their interest in the sustainable concept and their activity in favor of the environmental and the community: *Q1. Considering the Sustainable Development concept as an approach to pursue economic, environmental, and social goals, is your association council interested in promoting this concept within your members? Q2. Has your association initiated any kind of action among your members in favor of the environment during last year? Q3. Has your association initiated any kind of action among your members in favor of their own community during last year? Q4. Has your association joined any organization (government or ngo) in favor of the environment? Q5. Does your association contribute to any social group or organization in favor of less privileged communities? Q6. How many brands within your association are considered "sustainable" because of their product or service or because of their processes or because of their activities in favor of the environment or the community? Q7. Has your association recognized any brand for their interest and activity in pursuing environmental and/or social goals? Q8. Do you think your association council would be interested in participating in a global survey to find out the level of environmental and social activity being developed within its members′ networks?*

Responses. In both questionnaires, 22 national franchise associations replied within the deadline (51% of the total sent–another response was received after the deadline); 54% of respondents are located in developed countries, and 46% in emerging countries; even though the number of responses was the same, the participants were not the same; 63% of participants in the 2018 survey repeated in 2019, and 36% of participants were "new" for the 2019 survey (see Table 6.3). The respondents' size classification was: for both surveys (2018 and 2019): 81% were small associations with up to 300 brand members; for the 2018 survey, 9% were medium (between 301 and 400 brands); and for the 2019 survey, only 5% were medium and 5% were large (more than 1000 brands).

Table 6.3. National franchise associations participants in the 2018 and 2019 questionnaires.

Associations participating in the sustainable development questionnaire (2018–2019)		
	Year	**Country and name of the association participant**
1	2018–2019	Australia–Franchise Council of Australia
2	2019	Belarus–Franchise Association of Belarus
3	2018–2019	Brazil–Brazilian Franchising Association
4	2019	Britain–British Franchise Association
5	2018–2019	Colombia–Cámara Colombiana de Franquicia
6	2018–2019	Croatia–Croatian Franchise Association
7	2019	Czech Republic–Czech Franchise Association
8	2018–2019	France–French Franchise Federation
9	2018–2019	Greece–Franchise Association of Greece
10	2018–2019	Guatemala–Asociación Guatemalteca de Franquicias
11	2018–2019	Hungary–Hungarian Franchise Association
12	2019	Japan–Japan Franchise Association
13	2019	Lebanon–Lebanese Franchise Association
14	2019	México–Asociación Mexicana De Franquicias
15	2018–2019	Russia–Russian Franchise Association
16	2018–2019	Singapore–Franchising & Licensing Association (Singapore)
17	2019	South Africa–Franchise Association of South Africa
18	2018–2019	Spain–Spanish Franchise Association
19	2018–2019	Sweden- Swedish Franchise Association
20	2018–2019	The Netherlands–Dutch Franchise Association
21	2019	Turkey–Turkish Franchise Association
22	2018–2019	United States of America–International Franchise Association
23	2018	Finland–Finnish Franchising Association[2]
24	2018	New Zealand–Franchise Association of New Zealand
25	2018	Slovenia–Slovene Franchise Association
26	2018	India–Franchise Association of India
27	2018	Venezuela–Profranquicias (Venezuela)
28	2018	Argentina–Asociación Argentina de Marcas y Franquicias
29	2018	Poland–Polish Franchise Organization
30	2018	Taiwan–Association of Chain and Franchise Promotion Taiwan

Source: Own elaboration.

A frequency analysis was completed, and comments were grouped where they were considered to be more representative, according to the author. An additional analysis was completed to evaluate if there was any association between the responses and the economic level of the country.

[2] 2019 Finland's participation arrived after survey was closed and report had been issued.

Results and Analysis

The overall results provide sufficient evidence to show that the differences in size, location, country classification, and present or internal local interests of the participants' national franchise associations make it difficult to establish conclusive remarks to represent the national franchise associations as a whole on this matter. Here, and associating to concepts revealed in the 2017 MIT Sloan and BCG research, these two-year questionnaires to national franchise associations confirmed that sustainability interest among corporate executives (in this case, the associations' top management) remain just among a few and they are unevenly distributed across different countries (Kiron et al. 2017). The results of the 2018 and 2019 surveys are illustrated in Table 6.4.

Compared to the 2018 survey, the 2019 survey shows a slight increase in the interest, participation, and activities developed in the socio-environmental subject among the national franchise associations participants, as follows:

1. 73 percent of the respondents declared they have an interest or at least some interest in the sustainable development concept (versus 63% of 2018). This figure includes the 32% of the respondents who answered that "The council is interested and plans to promote the sustainability concept". Haanaes et al. (2011) define "embracers" and "laggards" companies regarding their commitment with sustainability. Here 32% of repondents represent the "embracers", who place high in their agenda the sustainability concept. On the other side, 41% represent the "laggards", cautiously adopting the concept.
2. 45 percent of respondents developed environmental initiatives with different levels of participation from their members (versus 21% of 2018). Here again, 32% had some favorable response from their members, matching the 32% of associations named as "embracers" in the paragraph above.
3. 46 percent of participants developed community initiatives with different levels of participation from their members (versus 54% of 2018). It is important to highlight that this figure includes 18% of "embracers" and 28% "laggards". Congruently, in this 46% of total replies, 70% of the responses are from the associations established in emerging countries, who have developed a higher commitment to their less favorable communities. This response matches with the 2017 MIT Sloan research, which states that the interest in the sustainability concept varies according to geographies, and in this case to the level of the country's economic development.
4. 23 percent is active in environmental organizations (versus 9% of 2018). Usually, this participation requires a monetary disbursement which might not be contemplated in the budget of the association. Again, this is a low percentage, illustrating the "embracers", who are "more willing" to make monetary investments for their sustainability commitment.
5. 50 percent of respondents participated in social organizations at different levels (versus 36% of 2018), matching the 46% of participants that developed

Table 6.4. Overall results for the entire sample (2018 and 2019).

Overall results for the entire sample (2018 and 2019)						
	2018*		**2019**		**Var.**	**Remarks**
1. Interest in sustainable development concept						For 2019, it shows a 19% increased interest in Sustainable Development concept. (Opportunity to work towards it.)
The council is interested and plans to promote it	5	21%	7	32%	11%	
There is some interest	8	42%	9	41%	8%	
No interest	7	21%	4	18%	–11%	
Do not know	2	8%	2	9%	1%	
Other	2	8%	0	0%	–8%	
2. Environmental initiatives						For 2019, it shows a 24% increase in the development of environmental initiatives, even with different levels of success.
No, we have not	17	79%	12	55%	–24%	
Yes, we did and 50% participated.	0	0%	2	9%	9%	
Yes, we did but was not successful.	2	8%	3	13%	5%	
Yes, we did and 30% participated.	3	13%	5	23%	10%	
3. Community initiatives						For 2019, it shows a 10% decrease in the development of community initiatives but at different levels of success.
No, we did not.	10	45%	12	55%	10%	
Yes, we did and 50% participated.	4	18%	4	18%	0%	
Yes, we did and 30% participated.	4	18%	5	23%	5%	
Yes, but it was not very successful.	4	18%	1	5%	–13%	
4. Participation in environmental organizations						For 2019, although there is an increase of 14% in the participation in environmental organizations, only 5 associations are active in these environmental groups.
Yes, we are active in environmental organizations	2	9%	5	23%	14%	
No, we have not	19	82%	17	77%	–5%	
Yes, but there was not much participation	2	9%	0	0	–9%	
5. Participation in social organizations						For 2019, there is a 23% increase in participation in social organizations, although only 11 associations participated in social organizations.
No, we did not contribute	14	64%	9	41%	–23%	
Yes, we permanently participate	3	14%	6	27%	13%	
Yes, we occasionally participate	5	22%	5	23%	1%	
Do not know	0	0%	2	9%	9%	
6. Sustainable brands						No major change is shown between 2018 and 2019. The only difference is the 15% decrease in the number of associations that replied they have no data.
001–25	12	50%	12	55%	5%	
26–50	1	4%	1	5%	1%	
51–75	1	4%	2	9%	5%	
more 150	0	0%	1	5%	5%	
No data	9	42%	6	27%	–15%	

Table 6.4 Contd. ...

...Table 6.4 Contd.

Overall results for the entire sample (2018 and 2019)						
	2018*		**2019**		**Var.**	**Remarks**
7. Recognition to sustainable brands						For 2019, it shows a 27% increased interest in recognizing their members for their efforts in the social-environmental issues.
No, we have not and is not a plan.	13	59%	7	32%	–27%	
Don't know	0	0%	1	5%	5%	
Yes, we did and plan to continue	4	18%	10	45%	27%	
No, we have not but plan to do so.	5	23%	4	18%	–5%	
8. Participation in future research						For 2019, no major change is shown, with half of the associations interested in evaluating/participating in future research on the subject.
Do not know.	2	9%	5	23%	–14%	
There is no interest.	6	27%	4	18%	–9%	
Interest for evaluating this.	7	32%	8	36%	4%	
Interest in participating	7	32%	4	18%	14%	
No answer.	0	0%	1	5%	5%	

* For 2018, the number of replies in all questions are not the same, as some respondents skipped questions and/or did not comment.
Source: Own elaboration.

social initiatives. This figure includes 27% as "embracers", who mention that they permanently participate with these organizations, and 23% represents the "laggards, who mention that they occasionally participate with these social organizations. Once again, this 50% of total figure includes 60% of associations from emerging countries, where there is a higher percentage of the less privileged population and requires a monetary disbursement from the association to help their immediate community.

6. 45 percent of respondents recognized sustainable brands (versus 18% of 2018). This involves a program recognition to the brands that initiate activities in favor of the environment or the community. An example of these recognitions is The Sustainability, Environmental Achievement and Leadership Awards (SEAL) launched in 2017, because they believe that "environmental progress requires true leadership, leadership deserves recognition, and recognition is a form of accountability" (Seal Awards 2017). This item also presents that 18% of respondents plan to recognize sustainable efforts in the future. Here again, the 45% figure represents the "embracers" pursuading others interested in this sustainable commitment, and the 18% figure represents the "laggards", who are "cautious adopters".
7. 54 percent of respondents mentioned an interest in evaluating or participating in future research on the subject (versus 64% of 2018). This figure includes the "embracers" by 18% and the "laggards" by 36 percent.

Hence, the 2018 and 2019 questionnaires validated that, in 2019, there was a 19% improvement in the national franchise associations' respondents for the interest in the sustainability concept, representing 41% of respondents as “embracers” and 32% of respondents as “laggards”.

The questionnaires also confirmed that the interest in the sustainability concept is unevenly distributed geographically, as it displays information that there is a certain causal relationship between the degree of the national franchise associations’ interest in the socio-environmental sustainability concept, and the level of the economic development of the country in which they are located (developed or emerging country). Figures 6.3 and 6.4 present a synthesis of the 2018 and 2019 survey results of the total sample as well as those of the subsamples of developed and emerging countries.

In 2019, the national franchise associations in developed and emerging economies show a significant difference in the degree of interest for the sustainable development concept (90% interest in emerging countries versus 66% interest in developed countries). When asked about the environmental initiatives developed, the difference in the replies is almost double, 60% in emerging countries, versus 33% of developed countries. In the development of community initiatives, the difference is again more than double, with 70% of respondents in emerging countries which are active with their communities, versus 33% of respondents in developed countries. When asked for their participation in social organizations, again the answers showed an important difference with a positive answer from the 60% of participants of emerging countries versus 25% of developed countries.

Another difference was in their interest for future research—60% of respondents in emerging countries agreed to participate versus 42% of respondents from developed countries, who replied they have no interest in participating or do not know if they would participate.

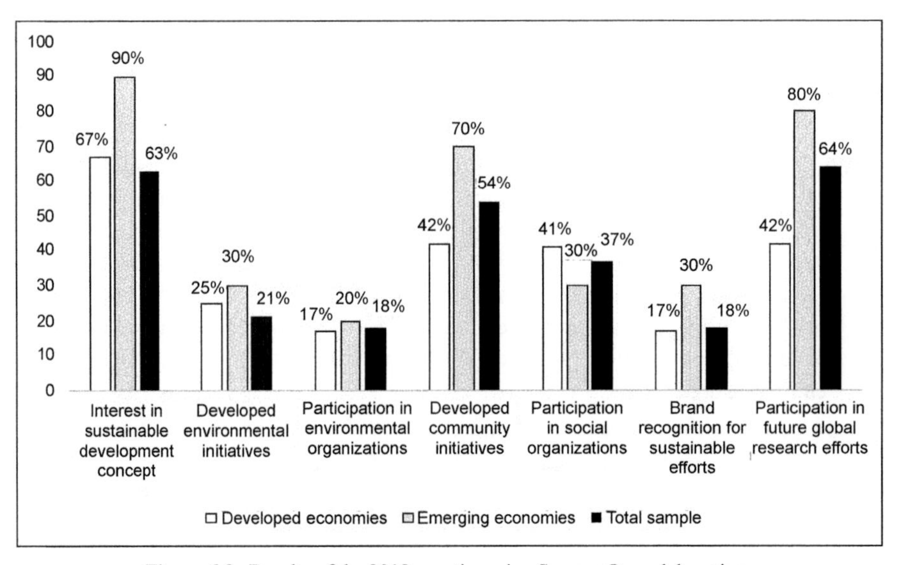

Figure 6.3. Results of the 2018 questionnaire. Source: Own elaboration.

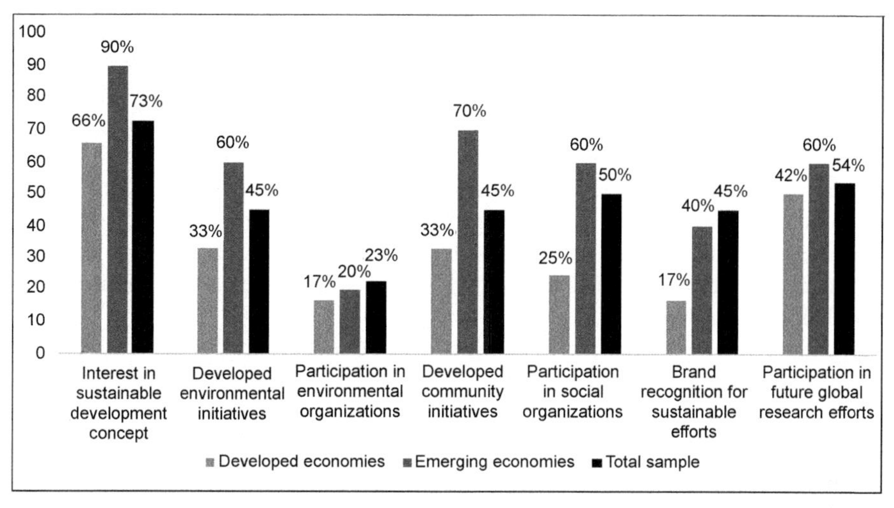

Figure 6.4. Results for the 2019 questionnaire. Source Own elaboration.

Survey Concluding Remarks

Survey Conclusion

The overall results provide sufficient evidence that show that differences in size, location, country classification, and internal local interests within the national franchise associations' participants make it difficult to establish conclusive remarks to represent the national franchise associations as a whole on the sustainability issue. Nevertheless, comparing against the 2018 survey, the 2019 survey shows a slight increase in the interest, participation, and initiatives developed for socio-environmental concerns. This slight increase highlights the participation of a handful of associations demonstrating their commitment towards sustainability, who are called the "embracers", according to the 2011 MIT Sloan and BCG research. Correspondingly, another handful of associations are cautious and slow in embracing this commitment; they are named the "laggards", according to the same research.

Another major finding was the relationship in the responses and the country's economic level where the association is located; this in turn, agrees with the 2017 MIT Sloan and BCG research supporting that the interest for sustainability is unevenly distributed across geographies and sectors.

General Conclusion

As mentioned before, the exploratory research objective was to find out the feasibility of further studies on the interest of business associations (representing a sector) to include the subject of socio-environmental sustainability in their business core to promote it among their members so as to increase its perception and consciousness to benefit their communities and the environment. This exploratory research illustrated the franchise sector, through the national franchise associations of 23 countries. The responses are considered interesting for further study regarding the importance of the participation of business associations to increase the knowledge and awareness in sustainability, to promote the concept, or to strengthen the commitment among their members (business enterprises), so as to develop and implement sustainable strategies and practices for a better environment and support to their communities.

Additional research should be followed to this exercise in other business sectors to validate the hypotheses that if business associations representing their sector promote the awareness and knowledge on the socio-environmental concept among their members, this concept would be embraced by their members and consequently, introduce changes in their business models and strategies in favor of the environment and their communities.

Chapter Concluding Remarks

Much has been said on the importance of innovation in the socio-environmentally sustainable concept in business organizations to provide real solutions to prevent, mitigate, and eliminate damages to the environment and the community due to business activities, which lead us to the new sustainable innovation concept. It was clearly stated that in order to achieve sustainability, innovations are tight to this aim, and at the same time, innovations also should include the sustainability concept.

The purpose of this chapter is to illustrate the importance of the most representative association (organization) of each sector in the diffusion and promotion of the sustainability concept among their members so as to provide knowledge, awareness, and consciousness to promote its inclusion in their members' business models.

The chapter presents advantages when participating in major business associations. Some advantages are: sharing current information, training and coaching, ethical business guidelines, cohesion against arbitrary decisions, future insights, among others.

Considering that business associations design business ethical guidelines for their sector and their members' behavior, sustainability fits perfectly into ethical guidelines as it is closely linked to how businesses create, produce, deliver value, and perform their activities. Consequently, these associations should accept their own accountability to protect the environment, health and safety, and, in general, the improvement of quality of life of their local communities and the environment. For this, business associations' top management need to design sustainability approaches to endorse, promote, evaluate, and recognize efforts from their members in favor of the environment and less favored communities.

Even though this two-year empirical research shows that in 2019, close to half of the associations' participants demonstrated, at least some interest in the sustainable theme, the author believes that if business associations embrace the sustainability concept, the effort would spread effectively among their members, and in turn, the concept could be included and implemented in their members' business organizations, because of the information and knowledge sharing on the subject as well as on the awareness developed. Likewise, these business associations would benefit from this effort, as they could gain recognition among their own members, and a competitive advantage within other business associations by becoming a reference in the social-environmental issue.

It is important to develop additional research on the kind and level of interest of the sustainability concept exerted by major business associations; as well as on their influence upon their members to introduce and develop sustainability strategies.

This could confirm or deny the hypothesis that "if major business organizations promote the sustainability concept among their members, then businesses would become conscious of their socio-environmental impact".

References

Adams, W.M. (2006). The Future of Sustainability: Re-Thinking Environment and Development in the Twenty-First Century. The World Conservation Union. Retrieved: 29 August 2018 http://cmsdata.iucn.org/downloads/iucn_future_of_sustanability.pdf.

Adner, R. (2016). Navigating the leadership challenges of innovation ecosystems. Retrieved: 24 March 2020. https://sloanreview.mit.edu/article.

Alarcón, N. (2015). La asociatividad como estrategia de desarrollo competitivo para las PYMES. Pensamiento Republicano. Bogotá, Colombia, D.C. N° 2. Publicado 2016-08-11. ISSN 2145-4175 pp. 13–31.

Albornoz, M. (2009). Indicadores de innovación: las dificultades de un concepto en evolución. Revista Iberoamericana de Ciencia, Tecnología y Sociedad - CTS, vol. 5, núm. 13 noviembre 2009, pp. 9–25, Centro de Estudios sobre Ciencia, Desarrollo y Educación Superior. Buenos Aires, Argentina.

Barjak, F., Niedermann, A. and Perrett, P. (2013). The Need for Innovations in Business Models. Final Policy Brief (Deliverable 5). Version 2.4. Submission 6th December 2013. European Commission DG Research and Innovation.

Berns, M., Townsend, A., Khayat, Z., Balagopal, B., Reeves, M. and Kruschwitz, N. (2009). The business of sustainability. Imperatives, Advantages and Actions. Retrieved: 25 March 2020. https://image-src.bcg.com/Images/BCG_The_Business_of_Sustainability_Sep_09_tcm9-170158.pdf. https://sloanreview.mit.edu/projects/the-business-of-sustainability/.

Blanda, B. (2019). Is "EcosystemOps" Going to be a Thing? Retrieved: 5 May 2020.https://blog.crossbeam.com/is-ecosystemops-going-to-be-a-thing?utm_campaign=Content%20Engagement%20Drip&utm_source=hs_automation&utm_medium=email&utm_content=83349673&_hsenc=p2ANqtz-8MAlvBPZMm6riwCz95kbbOp4EBbdrKlwtqIcshjltYG9s0eUZw6R53S9AjR_XxfNrO2vIIhBpnIPRfT4vaFHaq1YX9k5gzmSBWSpcgxtSYNsSPFcw&_hsmi=83349673.

Bocken, N.M.P., Short, S.W., Rana, P. and Evans, S. (2014). A literature and practice review to develop sustainable business model archetypes. Journal of Cleaner Production 65: 42–56. Retrieved: 13 May 2017: https://www.sciencedirect.com/science/article/pii/S0959652613008032.

Carrillo-Hermosilla, J., Del Río, P., Kiefer, C. and Callealta, F. (2016). Hacia una mejor comprensión de la eco-innovación como motor de la competitividad sostenible. Economía Industrial 401: 31–40. Retrieved: 23 January 2018: https://www.researchgate.net/publication/311909308_Hacia_una_mejor_comprension_de_la_ecoinnovacion_como_motor_de_la_competitividad_sostenible.

Christensen, C. (2020). Disruption 2020: An Interview with Clayton M. Christensen (interviewed by Karen Dillon). Innovation series. Disruption 2020. What it will take to innovate and compete over the next decade. Brought by: Deloitte. Spring 2020. MIT Sloan Management Review. Special Collection.

Çörtoğlu, F. (2019). The impact of sustainable development on the integration of European Union Climate policy. Contemporary Research in Economics and Social Sciences 3(1): 196–216. Ankara Universiti, Tu.

Crow, D. and Tett, G. (2020). Banks deny failure to combat climate change. Financial Times dated 22 January, 2020.

Daily News. (2020). Moody's Rising sea level risk to ratings. 17 January 2020–08.

Edgecliffee-Johnson, A. and Sevastopulo, D. (2020). Davos Economic Forum. Financial Times dated 22 January 2020.

EOI. (2007). Escuela de Organización Industrial 2007. La Innovación como Herramienta de transformación empresarial. Madrid: Fundación EOI. Retrieved:13 May 2018: https://www.eoi.es/es/savia/publicaciones/19425/la-innovacion-comoherramienta-de transformacion-empresarial.

Fink, L. (2020). Sustainability as new standard for investing. Retrieved: 21 February 2020. https://corpgov.law.harvard.edu/2020/01/17/sustainability-as-new-standard-for-investing/.

Fuller, J., Jacobides, M. and Reeves, M. (2009). The myths and realities of business ecosystems. Retrieved: 30 April 2020. https://ecosystemaces.com/topics/the-myths-and-realities-of-business-ecosystems.

Geradts, T.H.J. and Bocken, N. (2018). Driving sustainability-oriented innovation. Retrieved: 2 March 2020. https://sloanreview.mit.edu.article.

Haanaes, K., Arthur, D., Balagopal, B., Teckkong, M., Reeves, M. and Kruschwitz, N. (2011). Sustainability: The 'embracers' seize advantage. Retrieved: 4 March 2020. https://sloanreview.mit.edu/projects.

IBM Global Business Services. (2006). Expanding the Innovation Horizon: The Global CEO Study 2006. Retrieved: 23 May 2018: file:///C:/Users/Araceli/Downloads/March%201%20IBM%20CEO%20Event%20Presentation.pdf.

Kiron, D., Unruh, G., Kruschwitz, N., Reeves, M., Rubei, H. and Meyer, A. (2017). Corporate sustainability at a crossroads. Progress toward our common future in uncertain times. Retrieved: 2 March 2020. https://www.bcg.com/en-es/capabilities/sustainability/corporate-sustainability-at-crossroads.aspx.

Kopp, C. (2019). Creative Destruction. Retrieved: 26 April 2020. https://www.investopedia.com/terms/c/creativedestruction.asp.

KPMG. (2017). Perspectivas de Alta Dirección en México 2017. Web site KPMG: Press. Retrieved: 24 January 2018: https://home.kpmg/mx/es/home/sala-de-prensa/press-releases/2017/03/perspectivas-de-la-altadireccion-en-mexico-2017.html.

KPMG. (2018). Desarrollo sostenible en México, 2018. Web site KPMG: Press. Retrieved: 24 October 2018: https://cdn2.hubspot.net/hubfs/2866478/Landings_Estudios/PDFs/Estudio%20Desarrollo%20Sostenible%20en%20Mexico%202018%20TGP%20SC.pdf.

Layrisse, F. and Madero, S. (2018). ¿Sostenibilidad e Innovación, juntos y revueltos? Capítulo III – Temas Sustentables. pp. 283–305. *In*: Gomez, J.G.I., Sánchez, J.E. and Villareal, F.M. (eds.). Laglobalización ante los retos de sustentabilidad, Económico-Financiero y Organizacionales. Editorial Universidad Juárez del Estado de Durango, Mexico. ISBN: 978-607503-211-5.

Leff, E. (2008). Enrique Leff: Discursos sustentables. Editorial Siglo XXI, México, 2008, 272 p. Primera Edición. ISBN 978-607-3-00047-5.

Liendo, M. and Martínez, A. (2001). Asociatividad. Una alternativa para el desarrollo y crecimiento de las Pymes. Universidad Nacional de Rosario, Instituto de investigaciones económicas. Retrieved: 14 August 2019. https://www.fcecon.unr.edu.ar/web/sites/default/files/u16/Decimocuartas/Liendo,%20Martinez_asociatividad.pdf.

Lockett, H. and Geordiadis, P. (2020). Stock markets suffer coronavirus jitters. Financial Time International dated 22 January 2020, London, UK.

Lundvall, B. (1992). National Systems of Innovation: Towards a Theory of Innovation and Interactive Learning. Pinter Publishers, 1992. Universidad de California. ISBN1855670631, 9781855670631, Num. 2, 342 p.

Márquez, J. (2010). Innovación en modelos de negocio: La metodología de Osterwalder en la práctica. Revista MBA EAFIT, 30–47, sin datos. Retrieved: 24 January 2020. http://www.eafit.edu.co/revistas/revistamba/documents/innovacion-modelo-negocio.pdf.

Mendoza, H. (2011). La asociatividad empresarial, una estrategia para lograr competitividad. Estudio de caso. Instituto Politécnico Nacional, Escuela Superior de Comercio y Administración. Tepepan, México D.F.

Mintzberg, H., Quinn, J.B. and Voyer, J. (1997). El Proceso Estratégico: Conceptos, Contextos y Casos. Ed. Breve. Naucalpan De Juárez: Prentice Hall Hispanoamericana, 1997. 641 p. ISBN 9789688808290.

Montoya, O. (2004). Schumpeter, innovación y determinismo tecnológico. Scientia Et Technica, Año X (25): 209–213. ISSN: 0122-1701. https://es.scribd.com/doc/128233896/SCHUMPETER-INNOVACION-Y-DETERMINISMO-TECNOLOGICO.

Navarro, F. (2012). Responsabilidad social corporativa. Teoría y práctica. Barcelona: ESIC, 447 p.

OCDE. (2012). La estrategia de innovación de la OCDE. Empezar hoy el mañana. Edición de la OCDE Foro Consultivo Científico y Tecnológico 2012. Retrieved: 4 January 2016: http://www.foroconsultivo.org.mx/libros_editados/estrategia_innovacion_ocde.pdf.

OCDE/Eurostat. (2019). Oslo Manual 2018: Guidelines for Collecting, Reporting and Using Data on Innovation, 4th Edition. The measurement of scientific, technological, and innovation activities.

OECD Publishing, Paris/Eurostat, Luxemburg, Published on October 22, 2019. Retrieved: 20 January 2020. https://doi.org/10.1787/9789264304604-en.
Peraza, E.H. and Mendizábal, G.A. (2016). Sistemas Sectoriales de Innovación en España. Una tipología a partir de la encuesta sobre innovación en las empresas. Economía Industrial. 402: 117–127. Retrieved: 25 July 2017. https://www.mincotur.gob.es/Publicaciones/Publicacionesperiodicas/EconomiaIndustrial/Revista EconomiaIndustrial/402/PERAZA%20y%20MENDIZ%C3%81BAL.pdf.
Poliak, R. (2001). Asociatividad como grado de autonomía gerencial. Revista IDEA, mayo, 2001, Argentina.
Porter, M. and Van der Linde, C. (1995). Toward a new conception of the environment competitiveness relationship. The Journal of Economic Perspective 9(4): 97–118. Retrieved: 12 January 2017: https://www.jstor.org/stable/2138392?seq=1#page_scan_tab_contents.
Porter, M. (1998). Competitive Strategy, Creating and Sustaining Superior Performance. Free Press, 1998. First edition 1998. ISBN-10: 0684841460 ISBN-13: 978-0684841465.
Porter, M. and Kramer, M. (2011). Creating Shared Value. Harvard Business Review, January–February 2011 Issue https://hbr.org/2011/01/the-big-idea-creating-shared-value.
Regalado-Cerda, A. (2019a). Franchise associations interest in socio-environmental sustainability: An exploratory perspective (in press).
Regalado-Cerda, A. (2019b). Innovación sustentable en el sector de las franquicias en México. PhD. Thesis. Universidad Nacional Autónoma de México, Cd. De México, México.
Rodríguez, C. (2012). Sostenibilidad en las empresas. Retrieved: 12 November 2018. https://www.eoi.es/blogs/carollirenerodriguez/2012/05/20/sostenibilidad-en-las-empresas/.
Rosado, A., Paul, J. and Dikova, D. (2018). International franchising: A literature review and research agenda. Journal Business Research 85: 238–257. https://doi.org/10.1016/j.jbusres.2017.12.049.
Rost, K. (2011). The strength of strong ties in the creation of innovation. Research Policy 40(4): 588–604. Retrieved: 17 February 2016. https://www.sciencedirect.com/science/article/abs/pii/S0048733310002520.
Roy, S. and Purkayastha, S. (2007). Philosophy and strategy: Learnings from the great thinkers. The IUP Journal of Business Strategy. December, 2007.
Safran, J. (2019). We are the weather Saving the planet begins at breakfast. Penguin Random House, UK. ISBN: 978-0-241-40595-6.
Seal Awards. (2017). 2017 Business sustainability award winners. Retrieved 10 April 2020. https://sealawards.com/sustainability-award-2017/.
Searcy, C. (2018). Defining True Sustainability. Retrieved: 25 February 2020. https://sloanreview.mit.edu.article.
Thompson, A., Strickland III, A.J., Gamble, J.E. and Peteraf, H. (2012). Administración estratégica: teoría y casos. México: McGraw Hill. Edición 18va. ISBN:978-609-15-0757-0. EbookRetrieved: 20 January 2019. https://es.slideshare.net/MauraMaldonado1/ebook-administracin-estratgica-thompson-18va.
Tomlinson, B. (2018). How long-term investors influence corporate behavior. Retrieved: 23 February 2020. https://slowanreview.mit.edu.article.
UN. (1987). Our Common Future. Report of the World Commission on Environment and Development. UN Documents. Oslo, Norway. 20 March 1987.
UN. (2015). Sustainable Development Goals. Retrieved: 5 October 2017: https://sustainabledevelopment.un.org/index.php?menu=1300.
Vilaseca, B. (2008). Cómo crear una empresa sostenible. El País. Reportaje: Sección Gestión y Formación. January 2008. Retrieved: 29 April 2018: https://elpais.com/diario/2008/01/06/negocio/1199630853_850215.html.

CHAPTER 7

Customer Relationship Marketing (CRM) for the Design of Strategies in Digital Marketing

Raquel Ayestarán

Introduction

According to the traditional approach, marketing is conceived as the discipline consisting of the creation, communication, distribution, and exchange of relevant value propositions for consumers, customers, business partners, and society as a whole. According to the well-known definition of Kotler, these are processes aimed at satisfying the needs and desires of people through the creation, supply, and free exchange of products and services with high added value for them. This understanding, marketing is concretised in practice in a set of techniques that, in the form of a commercial strategy, serve the sales objectives of an organization according to its business model (Kotler and Kartajaya 2017).

As Stanton (1984) says, as it emanates from business activity, marketing deals with all aspects that help the sale: product planning, price allocation, and promotion and distribution in the market. From this perspective, the main purpose of marketing is to improve the marketing of a product (good or service), for which it is also essential to enter into a relationship with a certain group of people, potential consumers or customers, with the aim of discovering, identifying and, where appropriate, anticipating possible needs, patterns of behaviors and desires for satisfaction, chords to cultural changes and transferred to business. In order to achieve this goal, customer loyalty and brand positioning strategies are also ordered; they require a deep knowledge of the market and, ultimately, seek to arouse the recall or spontaneous evocation of the brand by inserting commercial references in the mind of the consumer (what is known as top of mind) according to Stanton (1984). In this

Universidad Francisco de Vitoria, Marketing Department, Spain; ayestaranraquel@gmail.com way, the recruitment of new customers is activated and loyalty is generated in those who already trusted in the brand when it comes to satisfying their needs for use and consumption. Finally, it is evident that marketing not only helps to place products on the market, but also constitutes an important factor of innovation and dynamisation, which allows companies to respond adequately at all times to the changing needs of potential recipients of their products (Kotler and Kartajaya 2017). The process is no longer unidirectional, nor is it limited to the management of the brand image or the logistics of the distribution, but it provides information on the characteristics of the good or service that is offered, on how they are perceived by the public, and on their real suitability to meet the needs of people (or other companies) in real time-aspects that, obviously, have a decisive influence on the long-term, can see how disruption trends and technology have broken out and affect the marketing sector. CRM or Customer Relationship *(Impact of customer relationship management (CRM) on customer satisfaction and loyalty: A systematic review)* is a concept that is based on the philosophy of using a combination of customers and marketing for relationship building, and argued that developing a relationship with customers is the best way to gain their loyalty. Customer loyalty refers to a consideration paid to the amount of buying for a given trademark.

Every day, there's seemingly yet another *disruptive* trend that emerges out of nowhere which affects consumer behavior and the future of everything along with it. Many of you already follow some of the most notable trends disrupting markets today and I know you're devising new strategies as a result in order to compete in these ever-shifting markets Chowdhry (2020).

- Real Time
- Social Media- Mobile
- Sharing Economy
- Peer-to-Peer Economy
- Maker Economy
- Internet of Things
- Crowdfunding/Lending

This wheel of disruption (see Figure 7.1) keeps turning and the Butterfly Effect it unleashes with each revolution is forcing the creation of new agile models to stay current let alone get ahead (Kotler and Kartajaya 2017).

For the writing of this chapter, we have used different sources—papers, especially manuals, and interviews with experts. This chapter is divided in five sections. First this introduction, the second is methods and objectives, the third section describes concepts of marketing 4.0, the fourth section shows the concept of CRM and finalizes with conclusions.

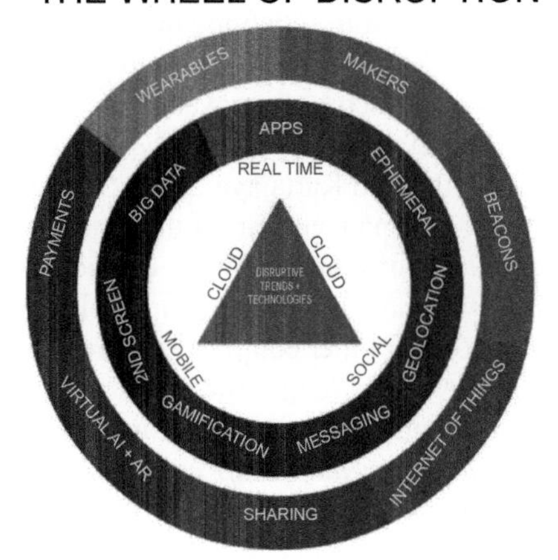

Within the Wheel of Disruption, the "Golden Triangle" is encircled by other emergent technologies and sectors affected by mobile, social, and real time, such as big data, geolocation, cloud and more.

Figure 7.1. The wheel of disruption. Source: Altimeter Group 2014 https://www.brian solis.com.

Methodology and Objectives

We have followed a process of collecting, analysing information, and drawing conclusions. These basic stages have been defined by the previously established objectives together with the contributions collected in the literature review. The collection of information has been carried out through literature review, especially interviews with professionals in the marketing sector; in-depth, open interviews have been conducted on the different key themes of this research. The reports from the marketing departments, as well as the attendance at seminars on CRM and ECRM, chosen to improve the qualitative data, have been of great interest. Studies and reports of results published on the reference websites of the mentioned sector have also been reviewed, as well as in books and minutes, official bulletins, and contrasted empirical research, which have provided us with relevant quantitative data to understand the evolution and the current situation of this strategy. The primary sources are the following:

- In-depth interviews, given to the author of this chapter, by professionals in the marketing sector.
- Information from the sector in more specific areas and cases: interviews, granted to obtain new, interesting data, unique for this research.
- Process and phases: In-depth interviews, with key questions that concern this investigation.

The secondary sources used in this study are the following:

- Documentary study: books and documentation mentioned.
- Interviews collected on websites in this sector, more current papers, this review being scarce due to the current nature of the topic to be discussed and direct sources.

- In-depth interview methodology: face-to-face open interview and the last one (face-to-face COVID 19) through Facetime. Panel: High consumption advertisers, Start Up's and marketing consultants. Universe: Marketing sector.
- Sample size and distribution: six interviews. Distribution of the sample in multiple stages stratified by type of business: large brands and SMEs. Sampling error: Negligible or nonexistent with 99% confidence level. Field work: carried out by the researcher, between September 2019 and April 2020.

Objectives are defined in relation to the Marketing Strategy and the CRM funnel phase. Thus, they are right in relation to the methodology we worked.

Concept of the Marketing 4.0

In today's daily life imposed by the market, the classic model just described is subject to constant revision and renewal. It has been several decades since brands began to dispense with marketing strategies aimed exclusively at selling (that is, directed above all to force the purchase decision, or simply not to generate stocks) to put the focus on the wishes and the needs of the people and the product that will be offered. New demands and new challenges appeared in this way, which roughly consist of:

- Identify, anticipate, and provide the needs and wishes of people with an objective profile.
- Define customer-oriented experiences and establish deep and lasting relationships between brand and user, based on loyalty, empathize, generate experiences without friction.
- Build and configure strategies that generate a tangible competitive advantage, while providing real value, which is also perceived in the market as such: innovation, commitment, and sustainability.
- Execute true and relevant experiences with the client, through the analysis of user behavior data, and monitoring of analytics, as a fundamental discipline, in marketing strategies.
- Concern about the brands in knowing what our client wants and making our USP (unique selling proposition) useful and valuable, improving it through active listening in all phases of brand-client relationship. This leads to the expansion of the traditional "four-fold" scheme (see Figure 7.2): the new person-centered marketing model gives rise to new processes . These processes generate dynamics that increase the benefits, to the extent that the strategies, having revised the necessary costs and resources, can better adjust to the real demand of the product and market conditions, depending on the profitability obtained. The most important change is that the brand proposes to people above all an experience, within a space that surrounds the product and that aims to reflect the values and personality of the brand itself. Presence and presentation are vital when integrating, in physical and online stores, the arguments aimed at provoking the purchase decision. In this way, both physical points of sale

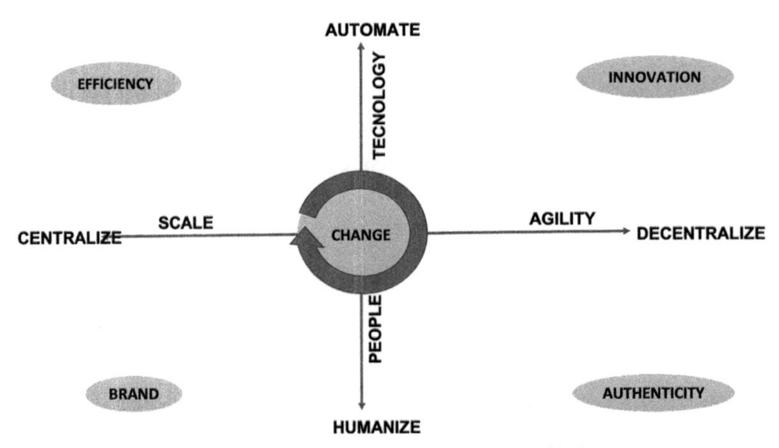

Figure 7.2. Automation marketing process. Source: chiefmatec.com.

and e-commerce platforms coexist today in a hybrid or mixed scenario, which is reflected in the notion of omnichannel. This scenario also highlights the experience and recommendation process that marketing has experienced in its stage 4.0 (Adigital 2015).

All this leads to conceive marketing as a discipline that includes a set of research, analysis, and marketing techniques that allow brands to measure, analyse and control the relationship phases, the experiences and the needs that consumers have (and how they can be met). We no longer speak exclusively of sales, but we incorporate a reference to the analysis of the habits and behaviors of people to put at their disposal what they require, adding value. In this way, it is essential to constantly attend to new consumer trends, such as the so-called "ROPO effect", an acronym that refers to the maximum Research Online, Purchase Offline; and that reflects a growing reality in society: consumers who seek information online, but who make purchases at physical points of sale (in fact, each time target audiences move more in their purchasing decisions as audiences to Mobile Marketing). In any case, it is about consolidating the digital transformation based on marketing. This philosophy advises that the Master Plan of the company be aligned with the Marketing Plan and the Commercial Plan; communication that must take into account the reality of an integrated and connected world. Kotler speaks in this sense of the "effect of collaboration", where companies treat consumers as human beings in all its dimensions: body, mind, spirit, and heart. A global, holistic, collaborative, integrated, and 360° environment is generated, in which the values, principles, and transparency of the companies are the factors that matter most to all citizens; and in which, for their part, brands establish contact points with them, which allow them to obtain in-depth information about the human process or the process that matters most to them: the one that starts with an interest in a good or service and ends with a decision of purchase or contracting in real time (what is known as customer journey, in all its phases of relationship and itineraries between both). Precisely this factor is the key that best reflects the displacement that has occurred in the position of dominance: power is now held by the client, the person; and it is the strategies of the brands that must provide real value, adapting and, insofar as possible, anticipating their wishes and needs (Aguado and García 2009).

In terms of marketing intelligence (in its two essential dimensions: strategic and operational), and also of commercial communication, the new central axis coincides, therefore, with the figure of the client. Currently, there are sophisticated market research and analysis techniques that allow targeting target audiences or, more recently, registered users or qualified data. Thanks to digital tools, databases are generated whose adequate treatment allows a greater personalization and interaction of companies with their stakeholders—what is known as CRM or Customer Relationship Management. At the same time, digital marketing campaigns are optimized in real time (real-time bidding or RTB) with the integration of algorithms, giving way to new professional profiles, such as brokers or mathematicians, in the purchase and sale of digital advertising spaces. All this is within the space of interrelation that is the Internet, where competitive advantages (such as those provided by logistics) or added value (for example, of total quality) point to the leadership of a collaborative economy. In this context, "relationship marketing" or "experiential marketing" is becoming increasingly important; and it has in its innovative or experimental version, the customer insights, the ideal test bench. In view of all the above, it is undoubted that, as an essential piece in a free market economy, marketing has existed, exists, and will exist to the extent that companies need information with a commercial service purpose, which allows them to adapt both the preferences of the recipients of their products and the circumstances of the business at all times. The market imposes its demands to the point that, today, the marketing professional is expected to know how to select and collect relevant information about consumer habits and customs, and then analyse it in large volumes and convert it into intelligent data (smart data). From this analysis will emerge the design of innovative strategies based on a certain knowledge of the business, which is a product of the processed data (business intelligence). To design these strategies, the marketing professional needs to possess a set of specific skills. Specifically, it has to be capable to:

1. Analyze adequately the target audience to which it is directed: technological segmentation.
2. Align objectives between corporate culture, brand values and attributes, business vision and value propositions, robust, true and relevant, to generate a real position, in line with its commitment to society and stakeholders.
3. Delimit the establishment of the strategy and its value proposition.
4. Define the competitive advantage to build relevant brands; that is to say, that by their authenticity they are true.
5. Align and execute tactics, integrate channels, and review compliance with the Action Plan (it is at this point where the deployment of the four legs occurs, always around the person).
6. Carry out the monitoring and control through the Balanced Scorecard, to measure and control the efficiency in the processes and the effectiveness in the results and

to be able to foresee, in each one of those phases of the funnel of conversion or funnel of sales, the life cycle of the brands in relation to the consumer and the market.

However, in environments like the current one, characterized by disruption and historical change, carrying out all of the above is particularly complex. To designate this situation, the acronym VUCA is often used, which refers to the volatility, uncertainty, complexity, and ambiguity present in the current market and society, and which lead to a certain requirement of transversally that is put in place, manifested especially in areas such as Digital Marketing. Regarding the context, it can be said that the market (especially in the digital field) is horizontal today, without borders. Brands thus face the need to adopt a "global" positioning (think in global terms but act at the same time in local terms), often situating different social and legal frameworks, which require a special adaptation effort. The professionals of the sector face in this way the need to adopt a strategic leadership approach, in which the following skills stand out:

- Capacity for influence: to make ideas, the vision of the company, become a reality, within matrix structures increasingly complex and changing, but also more dynamic.
- Dialogue and negotiation with stakeholders and generation of traction.
- Social impact, conceiving the company as a group of people; which requires leadership to develop them internally and generate an outward benefit.
- Ability to synthesize and focus: move quickly and make others move in the same direction, to execute with speed only what is really relevant.
- Entrepreneurial attitude: look for the form, tools, and resources to achieve business objectives in a creative and autonomous way.
- Ability to adopt tactical approaches, when circumstances require, without renouncing the long-term strategic vision, analysing indicators, and adjusting for new deviations. Finally, it cannot be ignored that, from the point of view of their level of development or evolution as a person, the public is not homogenous, but plural and changing; so that for any professional of the marketing these words appear like a challenge: "are the marks those that have to aspire to the values of the consumers, in place of being the consumers who aspire to the values of the marks".

Table 7.1 shows digital transformation of digital marketing from incomplete to complete transformation taking into account focus, objective, digital strategy, CEO involvement, involved areas, type of innovation, change management, talent management, and culture.

What is a CRM (Customer Relationship Marketing)?

Bindi Bhullar, senior analyst at Gartner Group, says about the CRM that: "It's a business strategy that places the customer as the heart of your company. Imagine what your company would be like if your client could redesign it to adapt it to their

Table 7.1. Digital transformation digital in marketing.

	Incomplete transformation	**Complete transformation**
Focus	Organization activity.	Organization activity.
Objectives	Cost savings, increase of sales and productivity. Short-termism.	New business model. Opportunities and threats from disruption.
Digital strategy	It doesn't exist.	Defined and communicated.
CEO involvement	Delegates to relevant areas.	Leader of the transformation.
Involved areas	CMO and CXO.	All.
Type of innovation	Technology and incremental innovation.	Disruptive.
Change management	Sporadic in some projects/areas.	Concerted action at all levels.
Talent management	To develop digital competencies.	To manage diversity. Collaborative and innovative competencies, without fear of taking risks. Attract talent.
Culture	Silo culture.	Innovation culture. Change in the company culture towards greater emotional intelligence.

Source: raquelayestaran.com.

needs. This is the company you need to be. "The business consultant Accenture defines it as: "The continuous process of identifying, directing, developing, tracking, selling, serving and improving high-value relationships with Clients, so as to generate sustained growth and profits". Relationship Marketing is a term that encompasses a wide range of concepts, conceives the business environment in a broad sense in which the existing relationships between the company and the players that surround it (suppliers, stakeholders, administration, etc.) are integrated. These relationships are formed by the phases within the process in which they are recognized under the terms of identifying, establishing, developing, maintaining, and finalising relationships between brands and people. During this process, the relationship can also be measured through the quality of the service, the value of the relationship, trust in the service or product, satisfaction, commitment, fidelity, essential phases so that the relationship perceived as a process is lasting in time and so that, as a result, a mutual benefit is achieved, such as repeat purchase, communication through techniques such as Buzz MK eWOM, the permanence of customers, the acquisition of these, and the improvement in recovery and understanding the process of listening to lost customers. The characteristics of the websites, the interaction with social media, are tools that serve to manage customer relationships in digital channels and help to measure the user's feeling and behavior, define the attitude of the user while measuring the quality of the relationship. Constant innovation in value propositions, with the help of information and communication technologies (ICT) help improve the experience in the relationship with users and measure the impact on the practice of marketing. The loyalty marketing phase, together with the strategy developed by some brands, seeks to maintain long-term relationships with the consumer. Loyalty is considered as one of the profitability assets of any company, hence the main function of a CRM, is to work the right offer to the client by the precise channel and at the right time.

Loyalty programs are the result of optimizing relationships with consumers of the product or service, through personalized offers and punctual discounts for frequent purchases, since consumers expect from the programs a rational benefit, exclusive discounts, personalized offers and ad hoc to their needs, and an emotional benefit through the true and relevant experiences of brand and user.

Table 7.2 describes the model CRM and decision support to design marketing strategies that includes support-Clients database (marketing campaigns, opportunity generation), Marketing–assignation of potential clients (opportunity generation, follow up of opportunities), Sales–invoice generation (Product delivery), and Orders–technical service (knowledge base).

In the 90s, marketing began an evolution due to the progression of technology. We can identify a greater focus on customer service, on the relationship with the customer, and on the customer's level of knowledge. The trail that users leave while browsing contains rich information that can be used effectively in the creation of relationships with brands. The navigation of a potential consumer offers information declared by, for example, completing a form on a web page information, or we can even infer their interests by analysing the history of navigation based on data. This information stored in the cookies allows us to offer an x-ray of the potential consumer based on their behavior. The analysis of the information helps through real time bidding, offering related advertising through retargeting, adapting the content of the media according to the interests of the reader or, even anticipating user searches, behavior prediction models. Digital media has become more important within relational marketing strategies thanks to the customer relationship and devices, and these have become the simplest channel to reach the customer. The irruption of social networks has changed our model of relationship in real life, and opens a door to reliable information about interests, influence, location of users. The analysis of these data enriches relational marketing plans, evolving the concept eCRM to social eCRM. What is the role of the consumer? Consumers assume the digital trail, as they increasingly have a greater knowledge of the information they offer, and expect an adaptation of commercial offers according to their interests. A priori, if we base marketing on the selection, use, and analysis of data, we will have greater efficiency in terms of costs, since we will better adapt the offer to consumer demand, and therefore communication will be more effective. In a context in which the consumer is increasingly informed, the fragmentation of markets, consumers have greater decision power thanks to an increasingly rich supply. Given this power of choice, loyalty will be the measure of success that will be taken into account as an indicator within a relational marketing plan, with the aim of maximizing the profitability of the value of its customers in the long term. Evolution is the origin of traditional

Table 7.2. CRM and decision support.

Marketing	Sales	Technical service	Supporting
Clients vs. opportunities	Invoices and Products sold	Clients' services	Database and Retargeting Real time bidding

marketing thinking, and it continues in relational marketing. How to work with each client is the challenge. If the average recruitment of a new consumer is higher than improving the loyalty of a current customer, brands will follow a different strategy to build loyalty to each person. The client has evolved to the e-client thanks to the digital transformation of the companies. You are trying to capture a more informed consumer than previously and you talk, interact, and demand in your relationship with brands (Aaker and Joachimsthaler 2005). The main characteristics of the e-client are:

- Use integrated channels to inform and communicate: omnichannel.
- Find opinions among your opinion leaders, or even other people.
- Share experiences with others and find people related to their interests.
- Look for an excellent shopping experience and want to participate in the process by giving your feedback.
- Technology barriers are broken and the digital channel is trusted as a sales and service channel. Within their relationship with e-clients, companies must base their strategy on transparency, information, multi-channel and user experience.

Parameters that are Needed in CRM

Through a meeting of the management team and that of the project, the following actions are carried out among others to implement CRM (see Table 7.3).

Conclusions

The cost of winning a new customer is much higher than the cost of retaining one. A satisfied consumer is one who is willing to repeat the purchase, recommend the brand, increase their expenditure or the frequency of the purchase, or even collaborate with the brand by offering their opinion to improve the product or service. In this sense, the main focus of a relationship-marketing plan is loyalty, identifying who the most profitable customers are in order to keep them, and then increasing revenues through the different actions that can be carried out. Customer data provides the information that allows the company to build maximum relevance with the consumer, enabling appropriate communication in terms of the message, context, form, and location. Data-driven marketing is based on data to define the strategy and is intrinsically related to CRM for the numerous reasons commented above. Thanks to the deep knowledge of the consumer and their needs, we can both improve the relevance of the brands through the content within the loyalty program, as well as align better to the business objectives. Through the data and records that it leaves, we are able to design brand strategies that align business objectives with the client's needs.

It is necessary to highlight that although a company undertakes a CRM strategy, this does not guarantee that immediately or even in the long term you will start to obtain higher revenues from your customers. For this to happen, CRM requires being part of the organizational culture and the acceptance of the clients involved in the process.

Table 7.3. Actions needed to implement CRM.

<table>
<tr><td>Show and discuss the customer pyramid</td><td colspan="2">This will help to explain the need that the company has to apply a CRM strategy.</td></tr>
<tr><td>Specify the objectives</td><td colspan="2">To define those objectives that the company wants to reach applying CRM to loyal clients.</td></tr>
<tr><td rowspan="5">Understand how the clients rate the company's value propositions</td><td colspan="2">It consists of defining which of the company's attributes are the ones most valued by its clients.</td></tr>
<tr><td colspan="2">Types of "value propositions"</td></tr>
<tr><td>Standard value propositions</td><td>Basic products or services that the company delivers to its clients.</td></tr>
<tr><td>"Standard +" value propositions</td><td>Aspects of the service which are complementary to the basic product.</td></tr>
<tr><td>"Standard ++" value propositions</td><td>Aspects of the relationships that add comfort and happiness to the client and create loyalty (and new purchases).</td></tr>
<tr><td>AdWords define the total costs of each current and potential client</td><td colspan="2">Define the total costs in the category of products and services of the company, and to estimate the changes in those costs in the near future.
When we talk about total cost, we mean the total sum that a client has spent or will spend on products and services in a specific period.</td></tr>
<tr><td>Determine the share of client</td><td colspan="2">The quantity of purchases that a consumer makes from the company, as a percentage of his total spending on the corresponding category. The company, with the objective of keeping or increasing that share, must analyse the factors that determines it, such as:
▪ The client's satisfaction level.
▪ The competitors, namely those ones that have the client share that the company doesn't have.
▪ Length, in years, of the relationship with the client.
▪ Client share from the previous year.</td></tr>
<tr><td rowspan="5">Review client information</td><td colspan="2">When executing a CRM strategy, the company needs to review the information that it needs on its client and where they are based (Aaker et al. 2004).</td></tr>
<tr><td>Basic Information</td><td>Personal data on current and potential clients, such as name, address, telephone number, fax number, email address, etc.</td></tr>
<tr><td>Information on Decision-making</td><td>It is interesting for the company to understand who plays the different roles in the decision-making process: the purchaser, the user, the final decision-maker, and those who influence the purchasing decision.</td></tr>
<tr><td>Financial Aspects of the Client</td><td>Costs should be identified that can be assigned to clients or client segments, so that client profitability can be estimated.</td></tr>
<tr><td>Client Behavior Variables</td><td>The unit of measurement upon which the client pyramid is built should be defined.
Generally, this unit is "sales revenue".</td></tr>
</table>

Source: raquelayestaran.es.

References

Aaker, D.A. and Joachimsthaler, E. (2005). Liderazgo de Marca. (V. e. Fons, Trad.) Barcelona, España: Deusto.

Aaker, J., Fournier, S. and Brasel, S.A. (Junio de 2004). When good brands do bad. Journal of Consumer Research 31: 1–16.

Adigital, A. p. (2015). Paid media use for the agencies. Adigital, AD digital agencies. Madrid: Adigital.

Aguado Guadalupe, G. and García García, A. (2009). Del Word-of-mouth al Marketing viral: Main trendin topicts about WOM and CRM. Comunicación y Hombre (5).

Altimeter Group. (2014). New Research on Enterprise Digital Transformation. https://www. briansolis. com/2014/04/brian-solis-altimeter-group-publish-new-research-enterprise-digital-transformation/.

Chowdhry, A (2020, April 06). Salesforce Hires Brian Solis as Global Innovation Evangelist. Brian Solis blog. https://www.briansolis.com.

Disruptive Technology is Disrupting Behavior. (November 6, 2015). The wheel of disruption. https:// www.briansolis.com/2015/11/disruptive-technology-disrupting-behavior/.

Kotler, P., Kartajaya, H. and Pearson. (2017). Fundamentals of Marketing. New Jersey. Wiley, J & Sons, Stanton. McGraw Hill. 2014. New Jersey.

Stanton, W. (1984). Fundamentals of Marketing. McGraw-Hill. New Jersey.

CHAPTER 8

CORSIA Evolution

A Global Scheme for a Sustainable Colombian Aviation Industry

*James Pérez-Morón** and *Lina Marrugo-Salas*

Introduction

Major events, such as 9/11, natural disasters, or severe acute respiratory syndrome-SARS, have led to periods where the aviation industry has seen reduced growth (Official Aviation Guide-OAG 2020a). The International Air Transport Association's (IATA) latest assessment states that the severe impact of the Corona Virus Disease-COVID 19 on air travel is even worse than that of the Global Financial Crisis (GFC), potentially changing the original CORSIA methodology (IATA 2020a), with a global air market "lockdown" during the second quarter, followed by an optimistic/rapid recovery of the aviation industry in the third, and continuing through the four quarter (IATA 2020a).

Nonetheless, the aviation industry grows permanently (Chapman 2007, Huang et al. 2020). One percent of the world's Gross Domestic Product (GDP) will be spent on air transport in 2020, or $908 billion in total (IATA 2019a). Between 2006 and 2016, worldwide demand for air passengers grew at an annual compound average rate of 3.7 percent (IATA 2016), 4.2 percent by 2019, and 4.1 percent by 2020 (Mazareanu 2020), and will double over the next 20 years (IATA 2019a). The number of aircraft seats has grown at an average annual rate of 3.6%, percent while available seat kilometers have grown by 4.6 percent (OAG 2020b, pp. 4). IATA (2018) predicts 8.2 billion people traveling by 2037, and by 2024, China will displace the United States as the world's largest aviation market. Becken and Carmignani (2020) question how realistic these estimations are, and consider them "extremely optimistic". See Air passenger market details in Table 8.1.

College of Business, Universidad Tecnológica de Bolívar, Campus Tecnológico - Parque Industrial y Tecnológico Carlos Vélez Pombo Km 1. Vía Turbaco, Cartagena, Colombia; lmarrugo@utb.edu.co
* Corresponding author: jperez@utb.edu.co

Table 8.1. Air passenger market details.

Market	Global Share	Revenue Passenger Kilometers (RPK)	Available Seat Kilometers (ASK)
TOTAL MARKET	**100%**	**4.5%**	**2.1%**
Africa	2.1%	5.4%	5.1%
Asia Pacific	34.7%	3.5%	2.5%
Europe	26.8%	2.5%	0.5%
Latin America	5.1%	1.5%	0.8%
Middle East	9.0%	5.9%	–0.4%
North America	22.3%	8.6%	4.4%

Source: Adapted by the author from IATA 2019b.

Academia, governments, and the industry have been reluctant to address aviation emissions openly. They do not want to point fingers at one of the favorite forms of passenger transportation, no matter how contaminating it may be (OAG 2020b, Peeters et al. 2018). However, different measures have been taken to break the relationship between the aviation industry and the environmental damage it causes (Chapman 2007, Lee et al. 2010), either by making it less polluting or by changing people's behavior (Banister and Hickman 2005, Marks 2019). Nowadays, the aviation industry is a key player in the world economy (Somerville 2003), with air freight being the fastest growing sector of that industry (Chapman 2007). Environmentally speaking, aviation pollutes both through carbon dioxide (CO_2) emissions, accounting for 2.4 percent of global CO_2 emissions (OAG 2020b), and with other greenhouse gas (GHG) emissions released into the atmosphere (Becken and Mackey 2017, Cairns and Newson 2006, O'Connell et al. 2019, Michaelowa 2016). IATA (2019c) highlights the aviation industry's three emissions goals for a more sustainable sector:

Table 8.2. Aviation industry emissions goals.

Goal 1	Improving fuel efficiency an average of 1.5% annually to 2020
Goal 2	Capping net emissions through carbon-neutral growth (CNG) from 2020
Goal 3	Cutting net carbon emissions in half by 2050 compared with 2005

Source: Adapted by the author from IATA 2019c.

Several other alternatives exist for mitigating the environmental impact of the aviation industry. One of these is taxation, which has been unsuccessful due to the difficulty of reaching an international agreement on the matter (Somerville 2003). Other options include policies to optimize air capacity (Chapman 2007), emissions permits (Chapman 2007) which provide the buyer the right to emit a certain amount of greenhouse gases into the atmosphere (Shi et al. 2020), or the Emissions Trading System–ETS, a market instrument for pollution control (Efthymiou and Papatheodorou 2019, Fang et al. 2019, Zhang et al. 2020, Zhang et al. 2020) that intends to minimize GHG costs (Becken and Mackey 2017, Somerville 2003). Sustainable Aviation Fuels (SAF) are also essential (IATA 2019b).

Aviation contributes to local emissions in multiple phases of flight (approach and takeoff nearby airports, during ground movements, and en route cruising). Greater emphasis is usually placed on CO_2 emissions because of their contribution to global warming, although other polluting or health-hazardous emissions exist, including nitrogen oxides and sulfur. Aviation has other harmful effects on the environment, including noise, discharges into water sources, and effects upon wildlife (The Civil Aviation Authority of Colombia-CAAC).

The two main emissions trading schemes are The European Union Emissions Trading System (EU ETS) (Pechstein et al. 2020) for aviation and the Carbon Offsetting and Reduction Scheme for International Aviation (CORSIA). In October 2016, the International Civil Aviation Organization (ICAO) adopted CORSIA, which will come into effect in 2021 to prevent carbon emissions from international aviation from exceeding 2020 emissions levels. The scheme requires participating airline operators to purchase carbon offset credits to compensate for any increase above 2019/2020 levels in their CO_2 emissions from international flights (Becken and Shuker 2019, Chao et al. 2019, Graver et al. 2019, Schneider et al. 2019a). IATA (2019c) has set the target of reducing net aviation carbon emissions to 50 percent of 2005 emissions levels by 2050.

Global aviation emissions are expected to reach 2.6 billion tons of carbon dioxide (tCO_2) by 2035. The countries responsible for 65 percent of those emissions are participating in the Pilot and First phases of CORSIA, some of which have the longest international aviation routes (see Table 8.3):

Table 8.3. Top ten global international routes.

Route		Flights	Miles
Newark	Singapore	728	8,277
Auckland	Doha	721	7,843
London Heathrow	Perth	730	7,829
Auckland	Dubai	710	7,664
Los Angeles	Singapore	1,034	7,611
Houston	Sidney	530	7.467
Dallas Fort Worth	Sidney	636	7,452
New York JFK	Manila	634	7,392
San Francisco	Singapore	2,482	7,330
Atlanta	Johannesburg	713	7,329

Source: Adapted by the author from OAG 2020a.

Latin America also has an active aviation sector. Its longest routes are Dubai to Sao Paulo and Santiago to Paris, and 80 percent of its international routes are to European airports (OAG 2020a). Table 8.4 contains the top ten longest international routes from or to Latin America.

Table 8.4. Top ten longest international routes from or to Latin America.

Route		Flights	Miles
Dubai	Sao Paulo Guarulhos	429	6,591
Paris	Santiago	377	6,298
Buenos Aires	Frankfurt	363	6,206
Mexico City	Tokyo Narita	730	6,074
Buenos Aires	Rome Flumicino	716	6,019
Paris	Buenos Aires	365	5,989
Madrid	Santiago	908	5,784
Sao Paulo Guarulhos	Istanbul Ataturk	364	5,694
Amsterdam	Lima	365	5,676
Barcelona	Buenos Aires	376	5,654

Source: Adapted by the author from OAG 2020a.

Colombia is only responsible for 0.35 percent of international aviation emissions (Climate Advisers 2017). In the Latin American region, Colombia currently has three rankings in the top ten busiest domestic routes: #1) Bogota to Medellin with 5,706,840 scheduled seats daily, #7) Bogota to Cali with 3,941,325, and #8) Bogota to Cartagena with 3,413,839 scheduled seats daily (OAG 2020a).

Colombia's aviation market is small compared to others in the region and globally, and despite Colombia's decision to not participate in the voluntary phases, CORSIA may represent a beneficial alternative to consider not only from a sustainability perspective for its aviation sector, but also for the ability to generate more than $322 million in additional private institutional investment at an estimated cost of $23 to $54 million to its aviation industry (Climate Adviser 2017, Hernandez and Rodriguez 2019, World Business Council for Sustainable Development. Mobility-WBCSD 2001).

There have been multiple review articles on CORSIA published in recent years (Abeyratne 2017, Chao et al. 2019, Erling 2018, Evans and Schroter 2017, Harper 2017, Lyle 2018, Maertens et al. 2017, 2019, Scheelhaase et al. 2018, Schneider et al. 2019a, b). Although informative, these studies have only reviewed a fraction of the literature. Thus, the first contribution of this study is to provide a more complete review of CORSIA by following the reporting checklist of the Preferred Reporting Items for Systematic Reviews & Meta-Analyses (PRISMA) (Liberati et al. 2009).

Another limitation of previous reviews is the exclusion of the Latin American aviation sector from their research. To the best of the author's knowledge, this is one of the first studies that addresses the potential lessons CORSIA can offer Latin American countries, specifically the Colombian aviation industry, and aims to contribute to filling this research gap.

The review conducted in this study is guided by the following main research question: 79 ICAO states (excluding Colombia) have volunteered to apply CORSIA in its early phases. In what ways can CORSIA make the Colombian aviation industry more sustainable?

This paper is organized as follows. The first section presents the method used to perform the literature review. The second section presents an overview of CORSIA-related research. The final section examines the Latin American and Colombian aviation market, its relation to CORSIA, and its challenges.

Methodology

Sample Selection

Based on previous reviews (Rosado-Serrano et al. 2018, Gancarczyk and Bohatkiewicz 2018, Gilal et al. 2018), the authors performed manual and electronic research in journals indexed in Scopus, and Clarivate Analytics' Web of Science (WoS). Later, the search was expanded to publications listed on Emerald, Science Direct, Wiley Online Library, Taylor & Francis, to ensure this study included the most recent documents on the subject published between 2016 and 2020. The starting point was set as 2016, since that was the year ICAO adopted CORSIA.

The journals containing the focal documents belonged to multiple domains, especially Environmental Science, Social Sciences, Agricultural and Biological Sciences, Business, Management, and Accounting and Engineering.

To select the final sample of articles, this research adopted a systematic four-step process. The first step was to identify keywords relevant to the subject matter. Besides "CORSIA", the search also included other keywords: Aviation, Gas Emissions, Sustainable Aviation, and Cross-border Emissions. Second, after reviewing with peers and experts in the field, this list only included the following relevant keywords: "CORSIA", "Sustainable Aviation", and "Aviation Gas Emissions". The search excluded "aviation" and "cross-border emissions" as search terms because both are too general. Both include other means of transport, and their research includes topics such as medicine. Given the aim of this study, this article consciously excluded documents that do not explicitly discuss CORSIA related concepts.

This study identified and combined keyword synonyms for a more comprehensive search. The search used both "free terms" and "index terms" funneled using the OR Boolean operator and the proximity operator W/n "within". The maximum distance between keywords was four words to ensure a closer relationship between the keywords selected in the final results. AND or AND NOT were included, since Scopus does not support using these operators as arguments to a proximity expression. This review was conducted following the PRISMA reporting checklist (Liberati et al. 2009).

Third, after selecting the search terms, the search returned 1,012 articles published in journals and containing each selected term in their title, abstract, and/or keywords. After eliminating duplicate articles, the list contained 607 unique focal articles. Fourth, both authors thoroughly reviewed the abstracts to validate the articles' relevance. This review led to the exclusion of articles unrelated to the CORSIA field of study. For instance, some articles deal with CORSIA in the health profession (e.g., Di Marco 2008), the arts and humanities (e.g., Eliav 2013), medicine (e.g., Milanesi and Sandrini 2010, Milanesi and Sandrini 2012), or genetics (Bodin et al. 2016, Rudall and Baterman

2002). This iterative process resulted in the final set of 23 focal documents (see Table 8.5). Figure 8.1 presents the results of the PRISMA analysis.

This study filtered the data in the final database using criteria, such as authors, journals, numbers of citations, references, and publishing dates. This study used the VOSviewer software to visualize the bibliometric maps.

Based on the initial sample of 23 articles, and for deeper analysis, three additional lists were created using Scopus features: (1) 981 references of the 23 focal documents, (2) 124 publications that cited the 23 focal documents, and (3) 927 references to these 23 publications. This study excluded the following documents: articles published in letters, editorials, note, incorrect affiliations, inaccessible abstracts, and wrong titles.

Table 8.5. Focal publications included in this review.

Outlets of focal publications	**Quantity**	**Reference**
Biotechnology for Biofuels	1	De Jong et al. 2017
Energy Policy	1	Staples et al. 2018
Journal of Air Transport Management	2	Scheelhaase et al. 2018, Winchester 2019
Climate Policy	1	Larsson et al. 2019
Environmental Development	1	Dube and Nhamo 2019
Transportation Research Record	1	Dray et al. 2018
Climate Law	2	Lyle 2018, Lecrerc 2019
Applied Energy	1	Chiaramanti and Goumas 2019
Journal of World Trade	1	Abeyratne 2017
Wirtschaftsdienst	1	Maertens et al. 2017
International Environmental Agreements: Politics, Law and Economics	1	Hoch et al. 2019
Sustainability	1	Maertens et al. 2019
Transportation Research Part D: Transport and Environment	1	Chao et al. 2019
Transportation Research Part B: Methodological	1	Zheng et al. 2019
Revista Brasileira de Política Internacional	1	Gonçalves and Anselmi 2019
Air and Space Law	1	Erlin 2018
WIT Transactions on Ecology and the Environment	1	Attanasio 2018
Airline Business	2	Evans and Schroter 2017, Harper 2017
Journal of Cleaner Production	1	Malins et al. 2020
Advances in Intelligent Systems and Computing	1	Aydogan and Zafeirakopoulos 2020
Total number of focal publications	**23**	

Source: Authors.

CARBON OFFSETTING AND REDUCTION SCHEME FOR INTERNATIONAL AVIATION (CORSIA) EVOLUTION: A GLOBAL SCHEME FOR A SUSTAINABLE COLOMBIAN AVIATION INDUSTRY

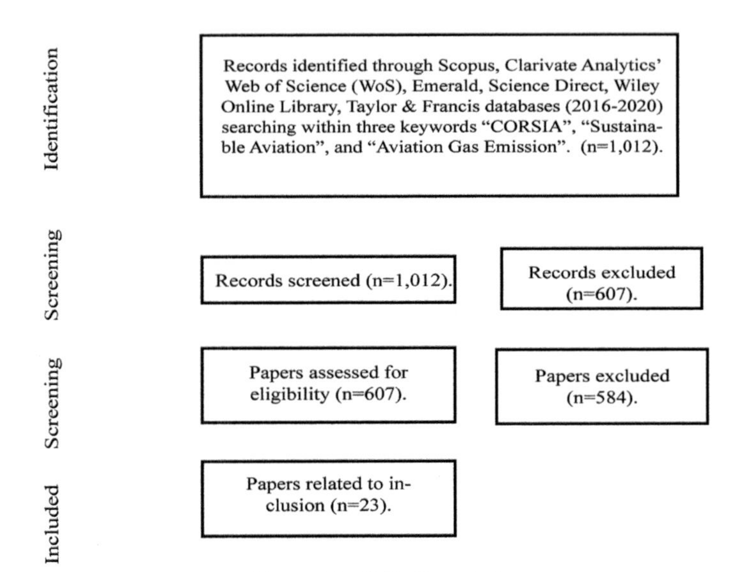

Figure 8.1. PRISMA analysis.

Results Analysis

Based on our filtered database of 23 documents, using criteria such as authors, journals, numbers of citations, references, and published dates, Table 8.6 presents the evolution of the selected publications from 2016 to 2020. The number of papers has increased over the years, from one publication in 2016 to ten in 2019. Only two articles have been published in 2020.

Articles published in scientific journals are the preferred document type (87%). Conference proceedings, reviews, and short surveys are the other types, representing 4% each.

Table 8.6. Published papers by year (2016–2020).

Year	Documents	% of 23
2016	0	0
2017	5	22
2018	6	26
2019	10	43
2020	2	9

Source: Authors.

Table 8.7. Document type.

Document type	Number of documents	% of 23
Article	20	87
Conference Paper	1	4
Review	1	4
Short Survey	1	4

Source: Authors.

The following table contains the top authors by number of publications. The most productive authors are Sven Maertens, Wolfgang Grimme, Martin Jung, and Janina Scheelhaase, with four publications each.

Table 8.8. Most productive authors.

Authors	Number of papers
Grimme, W.	4
Jung, M.	4
Maertens, S.	4
Scheelhaase, J.	4

Source: Authors.

Table 8.9 presents the 23 focal documents and journal titles organized by number of citations. Of these 23 documents, the 13 listed below have been cited 119 times. 2019 was the year where these documents were most cited, with 63 citations.

Data Visualization

This study used VOSviewer software for visualization, which allows developing the following bibliographic maps: co-occurrence, co-authorship, citation, and co-citation.

Co-occurrence

First, this study used VOSviewer to create a word co-occurrence map, including all keywords and full counting. From the 195 keywords extracted from all focal documents, the authors selected 43 that had two minimum occurrences. On the map (Figure 8.2), there are three clusters: (1) CORSIA in red, (2) climate change in blue, and (3) greenhouse emissions in green. The most used keywords are emissions control, carbon emissions, CORSIA, air transportation, climate change, aviation, carbon dioxide, greenhouse gases, and biofuels.

Co-authorship

This analysis reviews the links between the authors and their country distribution. For each of the 55 authors, the overall strength of co-authorship links with other

Table 8.9. "Extensively cited documents" (23).

Document title	Journal title	Total
		119
Life-cycle analysis of greenhouse gas emissions from renewable jet fuel production	Biotechnology for Biofuels	45
Aviation CO_2 emissions reductions from the use of alternative jet fuels.	Energy Policy	24
EU ETS versus CORSIA - A critical assessment of two approaches to limit air transport's CO_2 emissions by market-based measures	Journal of Air Transport Management	18
Climate change and the aviation sector: A focus on the Victoria Falls tourism route	Environmental Development	7
International and national climate policies for aviation: a review	Climate Policy	6
Impacts on the industrial-scale market deployment of advanced biofuels and recycled carbon fuels from the EU Renewable Energy Directive II	Applied Energy	4
The Global Potential for CO_2 Emissions Reduction from Jet Engine Passenger Aircraft. Transportation Research Record	Transportation Research Record	4
Beyond the ICAO's CORSIA: Towards a More Climatically Effective Strategy for Mitigation of Civil-Aviation Emissions	Climate Law	4
Carbon offsetting as a trade-related market-based measure for aircraft engine emissions	Journal of World Trade	3
How robust are reductions in modeled estimates from GTAP-BIO of the indirect land-use change induced by conventional biofuels?	Journal of Cleaner Production	1
Carbon offsetting and reduction scheme with sustainable aviation fuel options: Fleet-level carbon emissions impacts for U.S. airlines	Transportation Research Part D: Transport and Environment	1
Climate protection in air transport through market-based measures – from EU ETS to CORSIA, Klimaschutz im luftverkehr: Vom EU-emissionshandel zu CORSIA	Wirtschaftsdienst	1
CORSIA's struggle to offset doubts	Airline Business	1

Source: Authors.

authors will be calculated. The largest set of connected items consists of 55 authors, as can be seen in Figure 8.3.

Of 17 countries, Europe is the most productive region in terms of publications (Italy has six publications; Germany and the United Kingdom have five each). North America is the other key region with related publications: The United States has six publications and Canada has four. The top 5 citations per author country is as follows: United States (71), Italy (50), Netherlands (45), Canada (31), and Belgium (24) see Figure 8.4 Map of author countries.

The production of Latin America, Asia, and Africa is zero as of 2020, representing a challenge for academia in these countries, as it is important to be a part of this

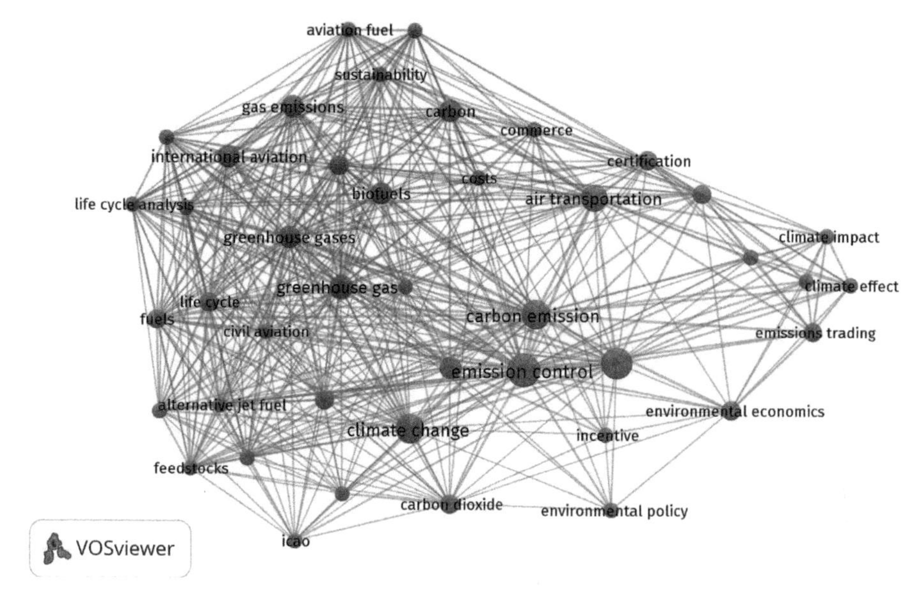

Figure 8.2. Word co-occurrence map (full counting).

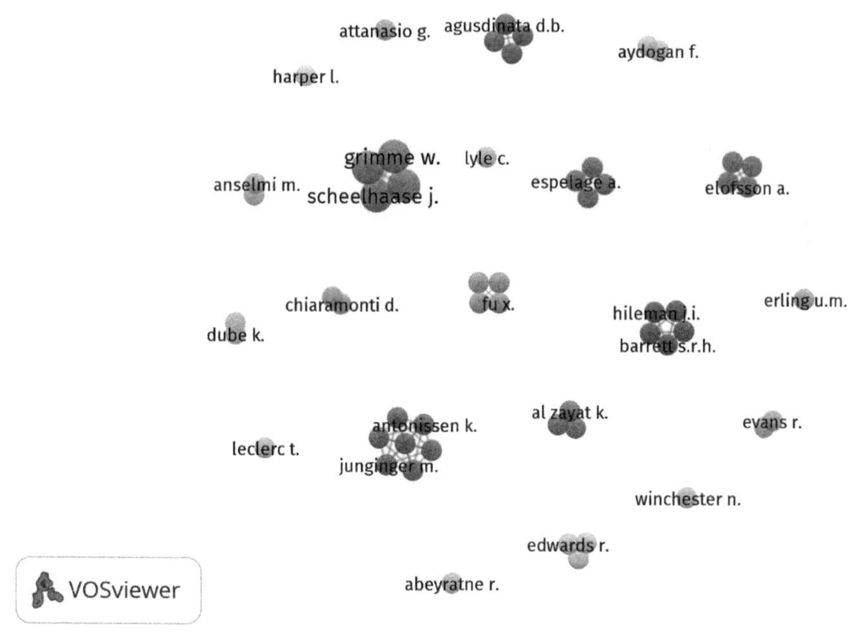

Figure 8.3. Map of authors.

global trend. To the best of our knowledge, this study is one of the first that covers the CORSIA scheme and contributes to reducing the intellectual gap in the field in Latin America. The minimum number of documents and the minimum number of citations per country have been set to one.

Figure 8.4. Map of author countries.

Citation

This section shows the connections and structure of most frequently cited focal documents. The minimum number of documents of a source has been set to one. VOSviewer examined 23 papers and selected 14 papers with at least one citation (Figure 8.5). This Figure 8.6 highlights the most cited authors and is consistent with the results shown in Table 8.9.

Figure 8.5. Document citations.

Deutsches Zentrum fur Luft- Und Raumfahrt has three authors who contributed publications, the Royal Botanic Gardens, Kew, Università degli Studi di Pavia, McNeese State University, and the Massachusetts Institute of Technology have two each. Starting with 55 authors, an author's minimum number of documents has been set to one, and VOSviewer selected 13 authors with the largest set of connected items. This figure is also consistent with Table 8.9.

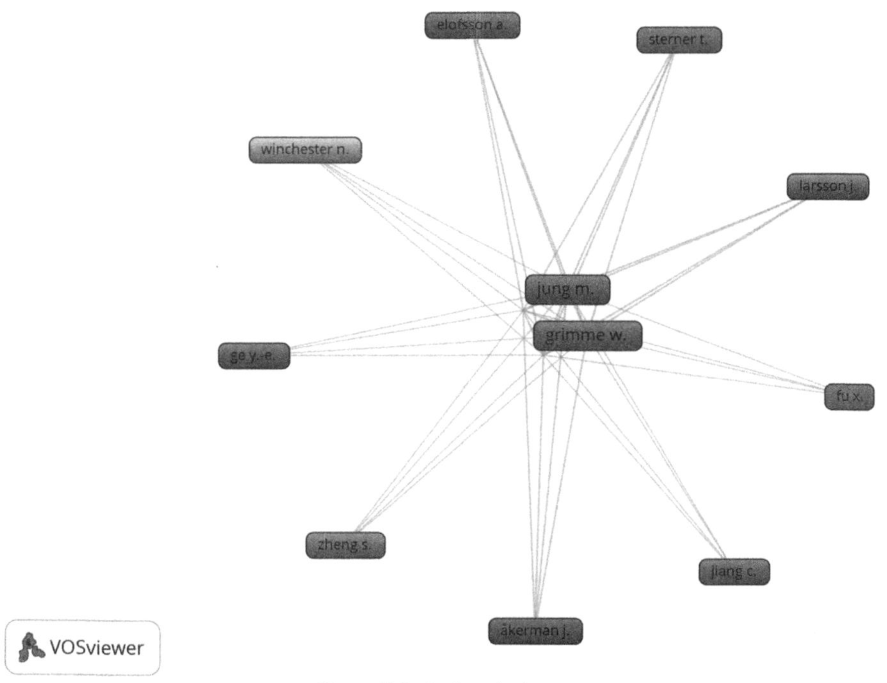

Figure 8.6. Author citations.

Co-citation

In this section, a new unit of analysis is selected, called "cited sources" to identify the most cited journals. The software extracted 396 of 458 cited sources with at least one citation, presenting the ones with the greatest total link strength (Figure 8.7).

What is CORSIA?

CORSIA is a global measure for the reduction of carbon emissions in international aeronautics implemented by the International Civil Aviation Organization. According to paragraph five of assembly A39-3, CORSIA is the first global market-based measure adopted by the sector. The objective pursued by the CORSIA initiative is to address, on an annual basis, all CO_2 emissions in international civil aviation that are increasing towards 2020 levels, considering special circumstances and respective

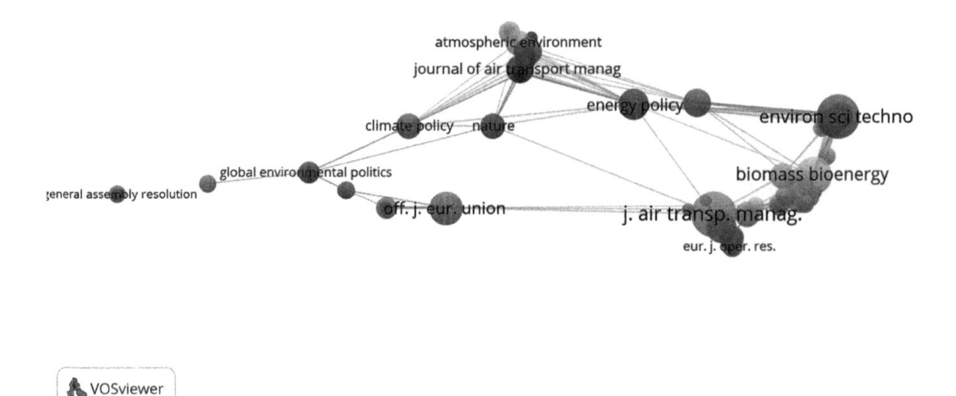

Figure 8.7. Co-citation of cited sources.

capabilities (ICAO 2019, IATA 2019a). By adopting this scheme, airlines will have to purchase carbon reductions to compensate for the growth in their CO_2 emissions. This compensation will be achieved by the implementation of carbon reduction projects mainly in developing countries and mostly linked to health and economic benefits, among others (IATA 2019a, 2019b).

The CORSIA scheme consists of a reference period and three implementation phases:

a) Reference period (2019–2020)

The CO_2 emitted by international aviation is averaged over the period from 2019 to 2020, and this is then used as a baseline to represent the growth of neutral carbon from 2020, which will reflect compensation requirements for the same year (share action operator to compensate for its emissions by financing a reduction in emissions).

CO_2 compensation requirements = Annual emissions of the operator x growth factor (ICAO 2019).

b) Pilot phase (2021–2023)

Voluntary participation by countries. This is the first cycle of measurement and verification of CO_2 emissions. The operator's compensation requirement applies to an operator's emissions covered by CORSIA in a specific year or an operator's emissions covered only in 2020 based on the calculation of compensation requirements.

c) The first phase (2024–2026)

73 states participate voluntarily, representing around 90 percent of international aeronautical activity. The operator's compensation requirement is based solely on the emissions of a specific year.

d) The second phase (2027–2035)

In the second phase, state participation is based on two categories of exemptions: aviation and socioeconomic criteria. Regarding aviation, all states that carry

Table 8.10. CORSIA pilot phase volunteer countries.

Country	Continent	Quantity	%
Bulgaria	Africa	5	7%
Burkina Faso	Africa		
Kenya	Africa		
Gabon	Africa		
Zambia	Africa		
Canada	America	6	9%
Costa Rica	America		
Guatemala	America		
Mexico	America		
United States	America		
El Salvador	America		
China	Asia	12	17%
Saudi Arabia	Asia		
Indonesia	Asia		
Israel	Asia		
Japan	Asia		
Malaysia	Asia		
Qatar	Asia		
Republic of Korea	Asia		
Singapore	Asia		
Thailand	Asia		
Turkey	Asia/Europe		
United Arab Emirates	Asia		
Albania	Europe	42	61%
Armenia	Europe		
Austria	Europe		
Azerbaijan	Europe		
Belgium	Europe		
Bosnia and Herzegovina	Europe		
Croatia	Europe		
Cyprus	Europe		
Czech Republic	Europe		
Denmark	Europe		
Estonia	Europe		

Table 8.10 Contd. ...

...Table 8.10 Contd.

Country	Continent	Quantity	%
Finland	Europe		
France	Europe		
Georgia	Europe		
Germany	Europe		
Greece	Europe		
Hungary	Europe		
Iceland	Europe		
Ireland	Europe		
Italy	Europe		
Latvia	Europe		
Lithuania	Europe		
Luxembourg	Europe		
Malta	Europe		
Monaco	Europe		
Montenegro	Europe		
Netherlands	Europe		
Norway	Europe		
Poland	Europe		
Portugal	Europe		
Republic of Moldova	Europe		
Romania	Europe		
San Marino	Europe		
Serbia	Europe		
Slovakia	Europe		
Slovenia	Europe		
Spain	Europe		
Sweden	Europe		
Switzerland	Europe		
The former Yugoslav Rep. of Macedonia	Europe		
Ukraine	Europe		
United Kingdom	Europe		
Australia	Oceanía	4	6%
Marshall Islands	Oceania		
New Zealand	Oceania		
Papua New Guinea	Oceania		

Source: Adapted by the author from ICAO 2019, Verifavia 2019.

out international civil aviation activities and meet the following two conditions participate:

- Their participation exceeds 0.5 percent of total Revenue Tonne Kilometers (RTK) of the activity in 2018.

  (Part. Ind. RTK = RTK of a State/Total RTK of all States)
- Their cumulative share of total activity is 90 percent (Individual RTK shares are classified from highest to lowest, and the shares are added in the same order until the value reaches 90 percent).

The following countries are stipulated as exempt based on socioeconomic criteria: "least developed countries (LDCs), small island developing States (SIDS), and landlocked developing countries (LLDC)" (ICAO 2019). Every state exempt from mandatory participation in the second phase of CORSIA, has full freedom to participate voluntarily since the objective is to cover at least 90 percent of international civil aviation activities. The wider the coverage, the greater the accuracy of emissions reductions. Table 8.11 lists all states that will participate in the second phase (voluntarily or otherwise).

Compliance with compensation requirements under CORSIA:

The operator notifies the consumption of sustainable aeronautical fuels during the compliance period. The state deducts the benefits for consumption of sustainable aeronautical fuels and reports the operator's final compensation requirements for the three-year compliance period. The operator purchases and cancels eligible emission units equivalent to its final compensation requirements for the compliance period. The operator sends a cancellation report of validated emission units to the state, which verifies the report and notifies ICAO (ICAO 2019).

Advantages of Participation in CORSIA

Environmental integrity is increased to the extent more states come together, and emissions coverage is extended. This means that, the greater the coverage, the more possibilities there will be for investment in projects that promote emissions reductions in developing countries. Voluntary participation by states in the pilot phase will give rise to priorities in terms of capacity building and assistance (ICAO 2019).

To reduce the administrative burden, CORSIA (both in compensation and MRV) will not be applied to aircraft operators that emit less than 10,000 metric tons of CO_2 per year, aircrafts with less than 5,700 kg of maximum take-off mass, or humanitarian, medical, and firefighting operations. IATA (2019) developed FRED+, a platform to support CO_2 emissions reporting for aircraft operators which are operational since January 2019, coinciding with the start of the CORSIA baseline period.

CORSIA Environmental Impact

A two-step estimate of the environmental impacts of CORSIA was carried out in 2018: the first step consisted of projecting the global increase in passenger traffic and CO_2 emissions for the period between 2017 and 2039. The tool used was 4D-Race,

Table 8.11. CORSIA second phase country participants.

Country	Continent	Quantity	%
Bulgaria	Africa	5	7%
Burkina Faso	Africa		
Kenya	Africa		
South Africa	Africa		
Zambia	Africa		
Brazil	America	7	10%
Canada	America		
Chile	America		
Costa Rica	America		
Guatemala	America		
Mexico	America		
United States	America		
China	Asia	14	20%
India	Asia		
Indonesia	Asia		
Israel	Asia		
Japan	Asia		
Malaysia	Asia		
Qatar	Asia		
Republic of Korea	Asia		
Saudi Arabia	Asia		
Singapore	Asia		
Thailand	Asia		
United Arab Emirates	Asia		
Russian Federation	Asia/Europe		
Turkey	Asia/Europe		
Albania	Europe	39	57%
Armenia	Europe		
Austria	Europe		
Azerbaijan	Europe		
Belgium	Europe		
Bosnia and Herzegovina	Europe		
Croatia	Europe		
Cyprus	Europe		
Czech Republic	Europe		
Denmark	Europe		
Estonia	Europe		

Table 8.11 Contd. ...

...Table 8.11 Contd.

Country	**Continent**	**Quantity**	**%**
Finland	Europe		
France	Europe		
Georgia	Europe		
Germany	Europe		
Hungary	Europe		
Iceland	Europe		
Ireland	Europe		
Italy	Europe		
Latvia	Europe		
Lithuania	Europe		
Luxembourg	Europe		
Malta	Europe		
Monaco	Europe		
Montenegro	Europe		
Netherlands	Europe		
Norway	Europe		
Poland	Europe		
Portugal	Europe		
Romania	Europe		
San Marino	Europe		
Serbia	Europe		
Slovakia	Europe		
Slovenia	Europe		
Spain	Europe		
Sweden	Europe		
Switzerland	Europe		
Ukraine	Europe		
United Kingdom	Europe		
Australia	Oceania	4	6%
Marshall Islands	Oceania		
New Zealand	Oceania		
Papua New Guinea	Oceania		

Source: Adapted by the authors from ICAO 2019, Verifavia 2019.

and the database was provided by Innovata, containing the global passenger flight plan for 2016. On average, a 3.2 percent increase in CO_2 emissions is expected, assuming an average increase of 1.2 percent pa in the efficiency of autonomous fuel, that is, from 734 million tons emitted in 2018 to 1,523 million in 2039. The second step was to use a tool developed by the *Deutsches Zentrum für Luft- und Raumfahrt*, the German Aerospace Center, to model the regulations under the scheme. Flight coverage between 67 states was used for the CORSIA simulations, and it was concluded that the initial CORSIA compensation rate would be relatively low, equal to 1.4 percent in 2021, but that its growth would be ongoing, such that, by 2030, 12 percent of passenger aviation emissions are expected to be compensated, reaching 18 percent by 2039 (Scheelhaase et al. 2018).

Because of its global nature, CORSIA can bring about a larger decrease in CO_2 than any national policy could achieve. According to forecasts, between 2021 and 2035, CORSIA will mitigate around 2.5 billion tons of CO_2 (164 million tons annually), allowing climate projects to be financed by up to US $40 billion in investments, and causing a US $12 increase in the price of carbon from 2021 to 2035 (IATA 2019c).

Double Counting

For more precise results in CO_2 emission reports, "double counting" must be avoided. Double counting implies counting the same emissions reduction more than once against climate mitigation objectives. If double counting is not regulated, greenhouse gas emissions could exceed the achievements reported by the countries involved. The carbon market includes three levels of action: (1) National or regional jurisdictions where policies allow companies to exchange emission permits or credits for reduced emissions according to a baseline, (2) Jurisdictions that link their national or regional policy instruments, allowing an international trade in permits or credits, (3) Article six of the Paris Agreement establishes a framework that allows countries to count these international transfers by demonstrating achievement of their objectives under the Paris Agreement (Schneider et al. 2019a, ICAO 2020).

Each participating country has individual carbon emission reduction objectives; therefore, the key to preventing double counting is that international transfers do not lead to higher overall emissions than individual targets. With this in mind, the Paris Agreement establishes "corresponding adjustments" as a form of accounting. In this manner, carbon emission negotiations are recorded in terms of the reduction, that is, the country that sells emission reductions must increase its own emissions, and the ones that buy said reductions, must subtract them.

To ensure compliance with this method, international transactions will be traced through electronic systems. Those involved must also submit periodic reports on their emissions and carbon market transactions, all subject to technical review. Politically, however, it is a challenge to solve the problem of double counting, since each country is different, with its own interests and promises to reduce emissions, so keeping accounting records of these transactions and ensuring their credibility is a complex task (Schneider et al. 2019b). In 2005, aviation contributed approximately 4.9 percent to the overall global radioactivity, and, with an increase in CO_2 emissions of 360 percent out to 2050, it is therefore very likely that the greatest demand for reduction of carbon emissions come from airlines (Markham et al. 2018, Warnecke et al. 2019).

Commercial, Legal, and Regulatory Challenges

The lack of legal guidance for applying CORSIA will represent a risk for regulators, who could find an aircraft owner responsible for non-compliance with CORSIA, but no clear action would be initiated by the contracting state against such a delinquent operator. There may also be legal and financial consequences if an operator does not pay off a sufficient number of allowable emission compensation units to cover any obligations arising after a declaration of insolvency (ICAO 2019).

According to ICAO (2019), establishing who is liable within the scheme when the operator of a flight has not been specified also represents a challenge for CORSIA implementation. Compliance with the scheme will fall to the owner of the aircraft in the following cases: (1) Where it has not been established who is the holder of the ICAO designator or the certificate of the air service operator, (2) In situations where the operator does not submit a monitoring plan and annual emissions report, and where, therefore, the contracting state cannot identify the operator or the person responsible for aviation emissions.

Considering that CORSIA is a scheme that is still in its initial stages, the United Nations Framework Convention on Climate Change (UNFCCC) has not yet developed the relevant accounting standards, and therefore, it falls to the ICAO to ensure that all CORSIA requirements match its regulations. To encourage country participation in the scheme, inclusive policies must be developed, including measures that consider the international aviation market, because every country has different policies, capacities, and structure (ICAO 2019). In their cutting-edge paper from 2019b, Schneider et al. summarized the key lessons learned from an assessment of the application of the first 14 carbon offset programs under CORSIA (see Table 8.12). They provided lessons for applicants and described how they satisfied ICAO Technical Advisory Body (TAB) requirements.

The Aviation Market in South America and the Caribbean

The airport sector represents an important element for the competitiveness of Latin America and the Caribbean. With improved air connections, transport costs can be minimized, foreign direct investment attracted, and trade improved overall (Economic Commission for Latin America and the Caribbean-ECLAC 2019).

IATA (2016) reveals the growth of international operations in South America since 2010 has been driven by Colombia (8 percent), Bolivia (6.1 percent), Chile (8.3 percent), and Peru (7 percent). This is due in large part to economic development in the region and the establishment of new airlines that have allowed opening up to external markets (Inter-American Development Bank 2020).

By 2018, international passenger traffic in the region had increased 6.9 percent, with a service capacity of 2.6 million flights and connections between 385 Latin American cities. Between 2008 and 2015, the average annual investment in the region was 0.05 percent of GDP (ECLAC 2019).

Table 8.12. The 14 program applicants.

Program	**Program Application Form**
American Carbon Registry	https://www.icao.int/environmental-protection/CORSIA/Documents/TAB/ACR_Programme_Application.pdf
British Columbia Offset Program	https://www.icao.int/environmental-protection/CORSIA/Documents/TAB/BC_Offset_Programme_Application.pdf
China GHG Voluntary Emission Reduction Program	https://www.icao.int/environmental-protection/CORSIA/Documents/TAB/CCER_Programme_Application.pdf
Clean Development Mechanism	https://www.icao.int/environmental-protection/CORSIA/Documents/TAB/CDM.pdf
Climate Action Reserve	https://www.icao.int/environmental-protection/CORSIA/Documents/TAB/CAR_Programme_Application.pdf
Forest Carbon Partnership Facility	https://www.icao.int/environmental-protection/CORSIA/Documents/TAB/FCPF_Programme_Application.pdf
Global Carbon Trust	https://www.icao.int/environmental-protection/CORSIA/Documents/TAB/Global_Carbon_Trust_Programme_Application.pdf
Gold Standard	https://www.icao.int/environmental-protection/CORSIA/Documents/TAB/Gold_Standard_Programme_Application.pdf
Myclimate	https://www.icao.int/environmental-protection/CORSIA/Documents/TAB/MyClimate_Programme_Application.pdf
Nori	https://www.icao.int/environmental-protection/CORSIA/Documents/TAB/Nori_Programme_Application.pdf
REDD.+	https://www.icao.int/environmental-protection/CORSIA/Documents/TAB/REDD.Plus_Programme%20Application.pdf
Thailand Greenhouse Gas Management Organization	https://www.icao.int/environmental-protection/CORSIA/Documents/TAB/TGO_Programme_Application.pdf
The State Forests of the Republic of Poland	https://www.icao.int/environmental-protection/CORSIA/Documents/TAB/State_Forests__Poland_Programme_Application.pdf
VCS Program (managed by Verra)	https://www.icao.int/environmental-protection/CORSIA/Documents/TAB/Verra_Programme_Application.pdf

Source: Authors.

Table 8.13 presents the top 30 out of 81 Latin American airports that move over one million passengers (ECLAC 2019).

Some highlights of Latin America's air routes can be seen in Figure 8.8.

Some Challenges of the Aeronautical Sector

The Latin American aviation sector is facing three main challenges. First, most of the region has extremely restrictive aero-commercial policies where the following shortcomings need to be addressed: free market access, pricing freedom, interventionism by the authorities, property liberalization and control, and stable regulations (ECLAC 2019).

Table 8.13. Busiest Latin American Airports-2019.

Item	Country	Airport	Passengers (in millions)
1	Mexico	Mexico Ciudad	44.5
2	Brazil	SP Guarulhos	37.0
3	**Colombia**	**Bogotá**	**31.0**
4	Mexico	Cancún	23.6
5	Chile	Santiago	21.4
6	Brazil	SP Congonhas	21.2
7	Peru	Lima	20.6
8	Brazil	Brasilia	16.5
9	Brazil	Rio-Galeao	15.9
10	Argentina	Aeroparque	13.3
11	Mexico	Guadalajara	12.7
12	Brazil	Belo Horizonte	9.8
13	Argentina	Ezeiza	9.7
14	Mexico	Monterrey	9.7
15	Brazil	Rio-Santos	9.0
16	Brazil	Campinas	8.8
17	Brazil	Porto Alegre	7.8
18	**Colombia**	**Antioquia**	**7.6**
19	Brazil	Recife	7.6
20	Brazil	Salvador	7.6
21	Rep. Dominicana	Punta Cana	7.3
22	Mexico	Tijuana	7.1
23	Brazil	Curitiba	6.5
24	Ecuador	Quito	6.5
25	Brazil	Fortaleza	5.8
26	**Colombia**	**Cali**	**5.0**
27	Costa Rica	San José	4.7
28	**Colombia**	**Cartagena**	**4.7**
29	Mexico	San José	4.7
30	Mexico	Puerto Vallarta	4.4
41	**Colombia**	**Barranquilla**	**2.6**
44	**Colombia**	**San Andres**	**2.3**
55	**Colombia**	**Santa Marta**	**1.7**
60	**Colombia**	**Bucaramanga**	**1.6**
62	**Colombia**	**Pereira**	**1.5**
78	**Colombia**	**Medellin**	**1.1**

Source: Adapted by the author from the InterAmerican Development Bank (IADB) 2020.

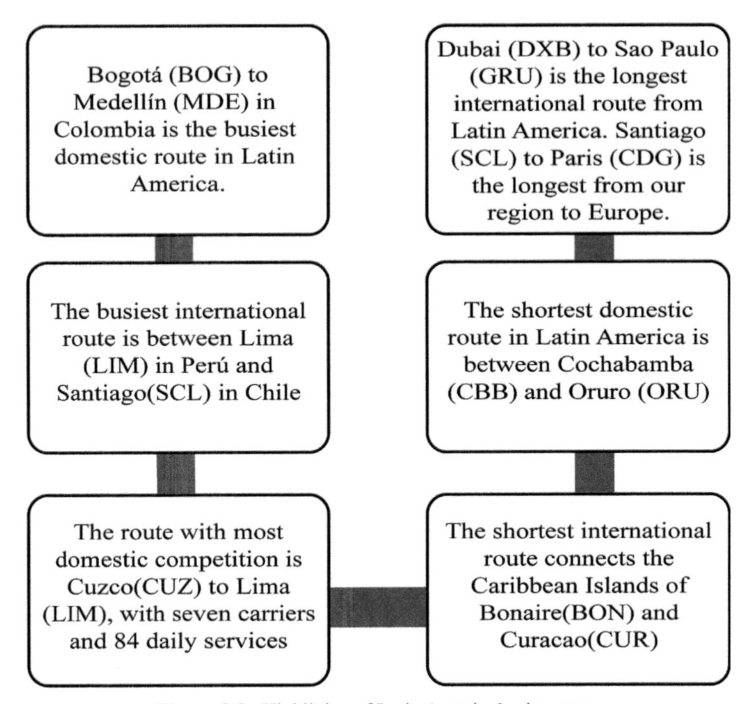

Figure 8.8. Highlights of Latin America's air routes.

Second, its infrastructure is incapable of addressing increasing levels of demand. The airport infrastructure is under great pressure due to progressive airspace liberalization and the region's economic growth. Third, there is a full breach of international safety standards imposed by the ICAO, reflected, on this point, by an increase in greenhouse gas emissions caused by air traffic in the region (ECLAC 2019, IADB 2020).

Recommendations

Perera (2009) and O'Connell et al. (2019) have proposed several ways to reduce GHG using technology to improve energy consumption, mainly with more efficient aircraft that use biofuels (Soria-Baledón and Kosoy 2018, Malins et al. 2020, Koistinen et al. 2019). However, a shortage of raw materials, a lack of infrastructure and a legal framework for refining, and a lack of financing are the main obstacles to the production and commercialization of these fuels (Chiaramonti and Goumas 2019, Deane and Pye 2018, IADB 2020). Other recommendations include:

- Using carbon footprints as an efficient alternative for reducing fuel emissions.
- The importance of creating new structures using composite materials to reduce aircraft weight together with aerodynamic improvements.

- Investigating the development of biofuels for aircraft using alternative advanced fuels obtained from plants and algae.
- Applying aerodynamic improvements. For example, the use of integrated winglets improves aerodynamic efficiency by between 3 and 5 percent.

CORSIA expects to provide two billion tons of investment-grade emissions reductions (2021–2035), and Colombia could benefit from this by entering the carbon market and connecting the above demand to domestic supply, including the Reducing Emissions from Deforestation and Forest Degradation (REDD+) project (Climate Advisers 2017, Hernandez and Rodriguez 2019), that focuses specifically on deforestation and forest degradation, and works through forest restoration.

Other projects that support sustainability in Colombia and are part of the local carbon market are: (1) the Clean Development Mechanism (see the work of Chen et al. 2020, Ahamad and Deepthi 2020, Kozlova et al. 2020), (2) GHG Emissions Compensation (see the work of Freebairn 2020, Gutiérrez and Lozano 2020), (3) voluntary market projects, and (4) Nationally Appropriate Mitigation Actions (NAMAs) (Climate Adviser 2017). Currently, in Colombia, carbon credit transactions are made up of emission mitigation projects using technologies, such as renewable energy installations, energy efficiency promotion and restoration, reforestation, and ecosystem conservation projects.

Colombia has decided not to participate in the voluntary phases of CORSIA. However, it has begun receiving assistance and training under a phase two ACT-CORSIA Buddy Partnership (ICAO 2020) with Italy as the donor state to prepare for CORSIA implementation in the territory (ICAO 2020). The CAAC has also committed to monitor CO_2 emissions from airlines.

Colombia expects its aviation traffic to grow, with its associated social and environmental impacts. The aeronautical infrastructure faces capacity shortages due to that growth (Balliauw and Onghena 2020, Jacquillat and Odoni 2018) together with multiple challenges, especially from security and environmental perspectives (Aeronautica Civil 2011).

Colombia's approach to tackling the above challenges is aligned with the Global Air Navigation Plan (GANP) and the Air Navigation Plan CAR/SAM, and their directions have been included in the National Air Navigation Plan (Plan de Navegación Aerea – PNA COL), which recognizes the importance of aeronautical infrastructure security and environmental sustainability (Aeronautica Civil 2011), and contributes to the ninth UN Sustainable Development Goal (9 SDG) "build resilient infrastructure, promote inclusive and sustainable industrialization, and foster innovation" through operational improvements (PNA COL 2020). Colombia's Air Navigation Plan also contributes to reaching the goals of the 2030 Aeronautical Strategic Plan – (Plan Estrategico Aeronautico – PEA 2030).

Colombia has 703 aerodromes (251 are public), 68 airports belonging to the CAAC, and 17 airport concessions. Figure 8.9 summarizes the Colombian airport concessions network (PNA COL 2017). Airport concessions have helped develop a more sophisticated airport infrastructure (Aeronautica Civil 2017) and seek to make airports more sustainable by reducing operating costs.

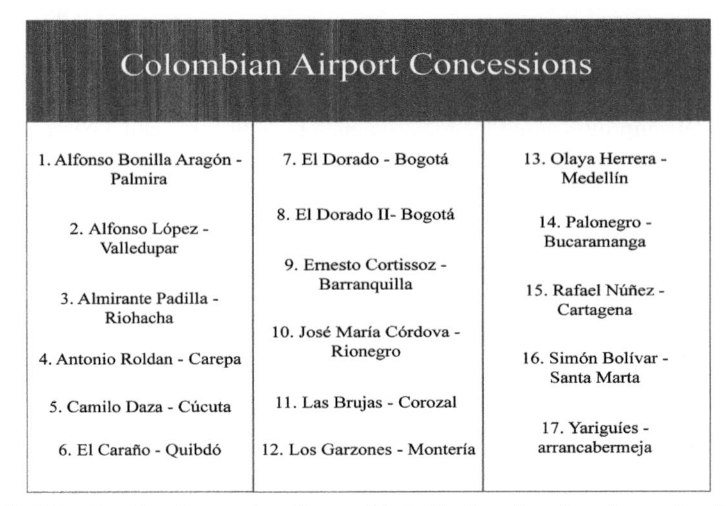

Colombian Airport Concessions		
1. Alfonso Bonilla Aragón - Palmira	7. El Dorado - Bogotá	13. Olaya Herrera - Medellín
2. Alfonso López - Valledupar	8. El Dorado II- Bogotá	14. Palonegro - Bucaramanga
3. Almirante Padilla - Riohacha	9. Ernesto Cortissoz - Barranquilla	15. Rafael Núñez - Cartagena
4. Antonio Roldan - Carepa	10. José María Córdova - Rionegro	16. Simón Bolívar - Santa Marta
5. Camilo Daza - Cúcuta	11. Las Brujas - Corozal	17. Yariguíes - arrancabermeja
6. El Caraño - Quibdó	12. Los Garzones - Montería	

Figure 8.9. Colombian airport concessions. Source: Adapted by the authors from Aeronautica Civil 2020.

ICAO (2019) provides a list of airplane operators attributed to states under CORSIA. Colombia reported the following airplane operator names (see Table 8.14).

The Colombian aviation industry has also implemented complementary strategies to become more sustainable, not only when airplanes are in the air (e.g., CORSIA), but also when they are on the ground. This is the case of the Airports Council International– ACI – Airport Carbon Accreditation program, which "provides a unique common framework and tool for active carbon management at airports with measurable results" (ACI 2015). In Colombia, only El Dorado International Airport is accredited at ACI level two/Reduction (airports that manage emissions intending to reduce their carbon footprint) ACI (ACI 2020). The Colombian aviation industry needs to strengthen its ecosystem and work together with other national institutions, such as the National Infrastructure Agency (Agencia Nacional de Infraestructura-ANI), the Ministry of the Environment, and the private sector.

Table 8.14. Colombian airplane operator names for CORSIA.

Airplane operator name		
Aerorepublica s.a.	Aerovias del Continente Americano s.a. "AVIANCA"/"avianca cargo"	Línea Aérea Carguera de Colombia s.a
Aerosucre s.a.	Lineas Aereas Suramericanas s.a.	Fast Colombia SAS
Aerovias de Integración Regional S.A.y/o LATAM Airlines Colombia		

Source: Adapted by the author from ICAO 2020.

Conclusions

The purpose of this study was to evaluate the literature on CORSIA using PRISMA and bibliometric analysis, as an initial step for research into the field. The authors organized scientific literature on CORSIA published between 2016 and 2020 and provided valuable insights.

The analysis provided a knowledge area map identifying word co-occurrence, links between authors and countries, citation and co-citation of the most cited documents. This study indicated that the most cited document, with 45 citations in Scopus over three years, was a publication by De Jong et al. (2017), and that the journals with the most cited papers were Airline Business, Climate Law, and Journal of Air Transport Management with two documents each. Forty three percent of the documents were published in 2019 (the most productive year), and 87 percent of the documents are articles.

The observation was that carbon emissions are the most common keyword and connect strongly with other keywords, such as CORSIA, Air Transportation, and Climate Change. Italy and the United States are the countries whose authors have the most publications with six each, followed by Germany and the United Kingdom, with five documents each.

One of the most remarkable results to emerge from the study is that Latin America, Asia, and Africa have zero publications related to CORSIA, which is a challenge for all authors in these regions to become a part of the CORSIA discussion. This study is one of the first steps towards enhancing our understanding of the CORSIA scheme for the Colombia aviation industry. The authors hope this research will be constructive in expanding the knowledge of CORSIA further, and to further this, the authors intend to compare state aviation industries that are already part of CORSIA. Hopefully, these studies will serve as a foundation for future studies on CORSIA.

Sustainability is vital for the Colombian Aviation Industry. This has been clearly outlined in the National Air Navigation Plan – (Plan de Navegación Aerea – PNA COL) and the 2030 Aeronautical Strategic Plan – (Plan Estrategico Aeronautico – PEA 2030) aligned with the Global Air Navigation Plan (GANP) and the CAR/SAM Air Navigation Plan. These plans also highlight the importance of considering local, regional, and national goals to promote the sustainability of the entire Colombian aviation sector, as CORSIA only considers international aviation.

Colombia is already receiving assistance and training under the ACT-CORSIA Buddy Partnership with Italy as a donor state, to prepare for CORSIA implementation in the territory (ICAO 2020). CORSIA will work together with a complementary program in Colombia's ACI, which will help to control and compensate GHE both in the air and on the ground. Other complementary projects that support sustainability in Colombia and are part of the local carbon market are: (1) the Clean Development Mechanism, (2) GHG Emissions Compensation, (3) voluntary market projects, and (4) Nationally Appropriate Mitigation Actions (NAMAs).

By joining CORSIA, Colombia will have the opportunity to measure, reduce, and compensate GHG emissions from aircraft. CORSIA will also allow Colombia to develop its aviation infrastructure according to international environmental standards, considering the implementation of additional ICAO programs that address new matters (noise, wildlife) not tackled by CORSIA or the ACI.

Airport concessions will help with the expansion of CORSIA and ACI programs since concessions have helped develop a more sophisticated airport infrastructure and seek more sustainable airports by reducing operating costs.

Further Recommendations for the Colombian Aviation Sector

All airports and airlines from the countries that have volunteered under CORSIA have realized that, between the technological and behavioral changes imposed by policies, they do have the ability to further reduce greenhouse gas and radiation emissions from the aviation sector. The truth is that there is no magical solution *per se,* but instead, a combination of measures can be used, including aviation fuel taxes, controls on low-cost airlines, better technology to, for example, change aircraft design to reduce fuel consumption and CO_2 emissions (Green 2003), regulations (via schemes such as CORSIA), and demand control (Scheelhaase et al. 2018). Alternative fuels with a low concentration of biodiesel will also allow people to "fly green" (Wardle 2003), together with changes to air traffic management (Somerville 2003).

These measures in isolation will have an insignificant effect on greenhouse emissions reductions, but, acting together, they may have the possibility to make the aviation sector more environmentally friendly.

Acknowledgments

The authors gratefully thank Dr Lambert Schneider from Oeko-Institut e.V. for his valuable support. Needless to say, this does not imply that he endorses the analysis in this publication.

References

Abeyratne, R. (2017). Carbon offsetting as a trade related market-based measure for aircraft engine emissions. Journal of World Trade 51: 425–444.

Aeronautica Civil. (2011). Aspectos institucionales de la organización de la Aeronáutica Civil en Colombia. NOTA DE ESTUDIO__Competencias Institucionales.pdf (aerocivil.gov.co)ttps://www.aerocivil.gov.co/aerocivil/foro2030/Documents/NOTA%20DE%20ESTUDIO__Competencias%20Institucionales.pdf.

Aeronautica Civil. (2017). Plan de Navegación Aérea para Colombia Volumen I. Requerimientos Operacionales. https://www.aerocivil.gov.co/servicios-a-la-navegacion/PublishingImages/planes-de-navegacion-aerea-para-colombia/BORRADOR%20PNA%20COL%20VOL%20I%20v08%203%20oct.pdf.

Ahamad, M.T. and Deepthi, M. (2020). Clean Development Mechanism and Green Economy. In Waste Management as Economic Industry Towards Circular Economy (pp. 49–58). Springer, Singapore.

AirportCarbonAccreditation.(2015).AchevingCarbonReduction.https://www.airportcarbonaccreditation.org/participants/latin-america-caribbean.html.

Airport Carbon Accreditation. (2020). Reducing Carbon $ Increasing Airport Sustainability. https://www.airportcarbonaccreditation.org/airport/4-levels-of-accreditation/introduction.html [retrieved on 30 March 2020].

Attanasio, G. (2018). Naples international Airport and Airport Carbon Accreditation (ACA). WIT Transactions on Ecology and the Environment 230: 465–474.

Aydogan, F. and Zafeirakopoulos, I.B. (2020). Leg base airline flight carbon emission performance assessment using fuzzy ANP (Conference paper). Advances in Intelligent Systems and Computing. 1029: 812–819. International Conference on Intelligent and Fuzzy Systems, INFUS 2019; Istanbul; Turkey; 23 July 2019 through 25 July 2019.

Balliauw, M. and Onghena, E. (2020). Expanding airport capacity of cities under uncertainty: Strategies to mitigate congestion. Journal of Air Transport Management 84(C).

Banister, D. and Hickman, R. (2005). Visioning and Backcasting for UK Transport Policy (VIBAT). New Horizons Research Program. Stage 1 Baseline Report. Department for Transport. The Bartlett School of Planning and Halcrow Group Ltd. Crown Copyright 2005, 81pp.

Becken, S. and Mackey, B. (2017). What role for offsetting aviation greenhouse gas emissions in a deep-cut carbon world? Journal of Air Transport Management 63: 71–83.

Becken, S. and Shuker, J. (2019). A framework to help destinations manage carbon risk from aviation emissions. Tourism Management 71: 294–304.

Becken, S. and Carmignani, F. (2020). Are the current expectations for growing air travel demand realistic? Annals of Tourism Research 80: 1–14.

Bodin, S., Kim, J.S. and Kim, J.H. (2016). Phylogenetic Inferences and the Evolution of Plastid DNA in Campynemataceae and the Mycoheterotrophic Corsia dispar D.L Jones & B. Gray. Plant Molecular Biology Reporter 34(1).

Cairns, S. and Newson, C. (2006). Predict and decide. Aviation, Climate Change and UK Policy, Environmental Change Institute, 122pp.

Chao, H., Agusdinata, D.B., DeLaurentis, D. and Stechel, E.B. (2019). Carbon offsetting and reduction scheme with sustainable aviation fuel options: Fleet-level carbon emissions impacts for U.S. airlines. Transportation Research Part D: Transport and Environment 75: 42–56.

Chapman, L. (2007). Transport and climate change: a review. Journal of Transport Geography 15: 354–367

Chen, H., Letmathe, P. and Soderstrom, N. (2020). Reporting Bias and Monitoring in Clean Development Mechanism Projects. Contemporary Accounting Research.

Chiaramonti, D. and Goumas, T. (2019). Impacts on industrial-scale market deployment of advanced biofuels and recycled carbon fuels from the EU Renewable Energy Directive II. Applied Energy 251.

Chiaramonti, D. (2019). Sustainable aviation fuels: the challenge of decarbonization. Energy Procedia 158: 1202–1207.

Climate Advisers. (2017). Vincular el mecanismo mundial de mercado de la OACI a la REDD+ en Colombia. https://www.climateadvisers.com/wp-content/uploads/2017/08/Brief-Linking-CORSIA-demand-to-REDD-in-Colombia-SPANISH.pdf.

Deane, J.P. and Pye, S. (2018). Europe's ambition for biofuels in aviation—A strategic review of challenges and opportunities. Energy Strategy Reviews 20: 1–5.

De Jong, S., Antonissen, K., Hoefnagels, R., Lonza, L., Wang, M., Faaij, A. and Junginger, M. (2017). Life-cycle analysis of greenhouse gas emissions from renewable jet fuel production. Biotechnology for Biofuels 10.

Dikova, D., Paul, J. and Rosado-Serrano, A. (2018). International franchising: A literature review and research agenda. Journal of Business Research 85: 238–257.

Di Marco. (2008). Hospital clown helps children cope with fears. 23(9), Gionale del Farmacista. ISSN: 03938476.

Dray, L.M., Schäfer, A.W. and Zayat, K.A. (2018). The global potential for CO_2 emissions reduction from jet engine passenger aircraft. Transportation Research Record: Journal of the Transportation Research Board.

Dube, K. and Nhamo, G. (2019). Climate change and the aviation sector: A focus on the Victoria Falls tourism route. Environmental Development 29: 5–15

ECLAC. (2019). Airport infrastructure in Latin America and the Caribbean. https://www.cepal.org/en/publications/44901-airport-infrastructure-latin-america-and-caribbean [retrieved on 30 March 2020].

Efthymiou, M. and Papatheodorou, A. (2019). EU emissions trading scheme in aviation: policy analysis and suggestions. Journal of Cleaner Production 117734.

Eliav, J. (2013). The Gun and Corsia of Early modern mediterranean galleys: Design issues and rationales. Mariners Mirror 99(3): 262–274.

Erling, U.M. (2018). How to reconcile the European Union Emissions Trading System (EU ETS) for aviation with the Carbon Offsetting and Reduction Scheme for International Aviation (CORSIA)? Air and Space Law 43: 371–386.

Evans, R. and Schroter, J. (2017). CORSIA's struggle to offset doubts (Short Survey). Airline Business 33: 48–49.

Fang, Z., Moolchandani, K., Chao, H. and DeLaurentis, D. (2019). A method for emission allowances allocation in air transportation systems from a system-of-systems perspective. Journal of Cleaner Production 226: 419–431.

Freebairn, J. (2020). A portfolio policy package to reduce greenhouse gas emissions. Atmosphere 11: 337.

Gancarczyk, M. and Bohatkiewicz, J. (2018). Research streams in cluster upgrading. A literature review. Journal of Entrepreneurship, Management, and Innovation (JEMI) 14: 17–42.

Gilal, F.G., Jian, Z., Paul, J. and Gilal, N.G. (2018). The role of self-determination theory in marketing science: An integrative review and agenda for marketing research. European Management Journal.

Gonçalves, V.K. and Anselmi, M. (2019). Climate governance and International Civil Aviation: Brazil's policy profile. Revista Brasileira de Política Internacional 62.

Graver, B., Zhang, K. and Rutherford, D. (2019). CO_2 emissions from commercial aviation, 2018. Working Paper 2019–16. The International Council on Clean Transportation. https://theicct.org/sites/default/files/publications/ICCT_CO2-commercl-aviation-2018_20190918.pdf [retrieved on 30 March 2020].

Green, J.E. (2003). Civil aviation and the environmental challenge. Aeronautical Journal 107: 281.

Gutiérrez, E. and Lozano, S. (2020). Efficiency performance of current Account-BoP flows in advanced world economies considering GHG emissions. Journal of Cleaner Production 120139.

Harper, L. (2017). CORSIA: Reducing expectations? Airline Business 33: 1619.

Hernandez, H. and Rodriguez, J. (2019). Participación de Colombia en el esquema de reducción y compensación de carbono para la aviación internacional (corsia) y el modelo de negocio verde. Master thesis. Universidad Distrital Francisco José de Caldas. http://repository.udistrital.edu.co/bitstream/11349/23102/6/HernandezMoralesHectorFelipe%20-%20RodriguezBuitragoJuanDavid.pdf [retrieved on 30 March 2020].

Hoch, S., Michaelowa, A., Espelage, A. and Weber, A.-K. (2019). Governing complexity: How can the interplay of multilateral environmental agreements be harnessed for effective international market-based climate policy instruments? International Environmental Agreements: Politics, Law and Economics.

Huang, F., Zhou, D., Hu, J. and Wang, Q. (2020). Integrated airline productivity performance evaluation with CO_2 emissions and flight delays. Journal of Air Transport Management 84.

Inter-American Development Bank. (2020). Perfil de las asociaciones público-privadas en aeropuertos de América Latina y el Caribe: principales cifras y tendencias del sector. https://publications.iadb.org/publications/spanish/document/Perfil_de_las_asociaciones_p%C3%BAblico-privadas_en_aeropuertos_de_Am%C3%A9rica_Latina_y_el_Caribe_Principales_cifras_y_tendencias_del_sector.pdf [retrieved on 30 March 2020].

IATA. (2016). 20 years passenger forecast. https://www.iata.org/contentassets/d4b60cffceeb4213bb5993d5fa2f358f/2016-10-18-02-es.pdf [retrieved on 30 March 2020].

IATA. (2018). IATA Forecast Predicts 8.2 billion Air Travelers in 2037. https://www.iata.org/contentassets/db9e20ee48174906aba13acb6ed35e19/2018-10-24-02-sp.pdf.

IATA. (2019a). Economic Performance of the Airline Industry. https://www.iata.org/en/iata-repository/publications/economic-reports/airline-industry-economic-performance---december-2019---report/ [retrieved on 30 March 2020].

IATA. (2019b). Air Passenger Market Analysis. https://www.iata.org/en/iata-repository/publications/economic-reports/air-passenger-monthly---dec-2019/ [retrieved on 30 March 2020].

IATA. (2019c). Annual Review 2019. https://annualreview.iata.org/?_ga=2.181191460.2080106556.1587872170-559893545.1587872169 [retrieved on 30 March 2020].

IATA. (2020a). IATA Economics' Chart of the Week Return to air travel expected to be slow. https://www.iata.org/en/iata-repository/publications/economic-reports/return-to-air-travel-expected-to-be-slow/ [retrieved on 30 March 2020].

IATA. (2020b). CORSIA and COVID-19. https://www.iata.org/en/policy/environment/corsia/ [retrieved on 30 March 2020].

ICAO. (2019). CORSIA Emissions Unit Eligibility Criteria. https://www.icao.int/environmental-protection/CORSIA/Documents/ICAO_Document_09.pdf [retrieved on 30 March 2020].

ICAO. (2020). CORSIA Buddy Partnerships. ICAO Environment. https://www.icao.int/environmental-protection/CORSIA/Pages/CORSIA-Buddy-Partnerships.aspx. https://www.icao.int/environmental-protection/CORSIA/Documents/CORSIA%20Aeroplane%20Operator%20to%20State%20Attributions_28May2019.pdf#search=colombia%20corsia [retrieved on 30 March 2020].

Jacquillat, A. and Odoni, A. (2018). A roadmap toward airport demand and capacity management. Transportation Research Part A: Policy and Practice, Elsevier 114(PA): 168–185.

Koistinen, K., Upham, P. and Bögel, P. (2019). Stakeholder signalling and strategic niche management: The case of aviation biokerosene. Journal of Cleaner Production.

Kozlova, E.P., Potashnik, Y.S., Artemyeva, M.V., Romanovskaya, E.V. and Andryashina, N.S. (2020). Formation of an effective mechanism for sustainable development of industrial enterprises. In Growth Poles of the Global Economy: Emergence, Changes and Future Perspectives 545–556, Cham: Springer.

Larsson, J., Elofsson, A., Sterner, T. and Åkerman, J. (2019). International and national climate policies for aviation: a review. Climate Policy 1–13.

Leclerc, T. (2019). A sectoral application of the polluter pays principle: lessons learned from the aviation sector. Climate Law 9(4): 303–325.

Lee, D.S., Pitari, G., Grewe, V., Gierens, K., Penner, J.E., Petzold, A. and Berntsen, T. (2010). Transport impacts on atmosphere and climate: Aviation. Atmospheric Environment 44(37): 4678–4734. doi:10.1016/j.atmosenv.2009.06.005.

Liberati, A., Altman, D., Tetzlaff, J., Mulrow, C., Gøtzsche, P., Ioannidis, J. and Moher, D. (2009). The PRISMA statement for reporting systematic reviews & meta-analyses of studies that evaluate healthcare interventions: Explanation & elaboration. PLoS Medicine 6(7): e1000100.

Lyle, C. (2018). Beyond the ICAO's Corsia: Towards a more climatically effective strategy for mitigation of civil-aviation emissions. Climate Law 8(1-2): 104–127. doi:10.1163/18786561-00801004.

Maertens, S., Scheelhaase, J., Grimme, W. and Jung, M. (2017). Klimaschutz im Luftverkehr: vom EU-Emissionshandel zu CORSIA. Wirtschaftsdienst 97(8): 588–595. doi:10.1007/s10273-017-2181-7.

Maertens, S., Scheelhaase, J., Grimme, W. and Jung, M. (2019). Options to continue the EU ETS for aviation in a CORSIA-World. Sustainability (Switzerland). Volume 11, Issue 20, 1 October 2019, Article number 5703. DOI: 10.3390/su11205703.

Malins, C., Plevin, R. and Edwards, R. (2020). How robust are reductions in modeled estimates from GTAP-BIO of the indirect land use change induced by conventional biofuels? Journal of Cleaner Production. Volume 258, Article number 120716 DOI: 10.1016/j.jclepro.2020.120716.

Markham, F., Young, M., Reis, A. and Higham, J. (2018). Does carbon pricing reduce air travel? Evidence from the Australian "Clean Energy Future" policy, July 2012 to June 2014. Journal of Transport Geography 70: 206–214. doi:10.1016/j.jtrangeo.2018.06.008.

Marks, P. (2019). Green sky thinking. New Scientist 241(3211): 32–36. doi:10.1016/s0262-4079(19) 30023-5.

Mazareanu, E. (2020). Global air traffic – annual growth of passenger demand 2006–2020. https://www.statista.com/statistics/193533/growth-of-global-air-traffic-passenger-demand/ [retrieved on 30 March 2020].

Michaelowa, A. (2016). Tackling CO_2 emissions from international aviation: challenges and opportunities generated by the market mechanism 'CORSIA'. Zurich Open Repository and Archive, University of Zurich. https://www.zora.uzh.ch/id/eprint/130452/1/AM2_corsia-airline_offsets-uae-eda11-16.pdf. https://doi.org/10.5167/uzh-130452 [retrieved on 30 March 2020].

Milanesi, P. and Sandrini, G. (2010). Philosophers on the wards: Clinical practice as a philosophical frontier. 19(3). Confinia Cephalalgica

Milanesi, P. and Sandrini, G. (2012). Project "Philosophy in the ward": How is the philosophical question in the context of the treatment of existential distress. 21(2). Confinia Cephalalgica

National Air Navegation Plan-"Plan de Navegación Aerea" (2020). 2030 Aeronautical Strategic Plan-" Plan Estrategico Aeronautico" (PEA 2030).

National Air Navegation Plan-"Plan de Navegación Aerea" (2020). http://www.aerocivil.gov.co/servicios-a-la-navegacion/Documentos%20vigentes/PNACOL%20Vol%20I%20v09%2019%20mar%20 2020.pdf [retrieved on 30 March 2020].

Official Aviation Guide – OAG. (2020a). How Green is your airline? The sustainability debate. https://www.oag.com/hubfs/free-reports/2020-reports/how-green-is-your-airline/how-green-is-your-airline.pdf?hsLang=en-gb [retrieved on 30 March 2020].

Official Aviation Guide – OAG. (2020b). Busiest Routes. https://www.oag.com/reports/busiest-routes-2020?hsCtaTracking=65c5d0e9-7928-40f7-b1d2-56e8ce2cfa38%7C9eb82cd8-0cc2-484c-b45d-b721d9cf87ec [retrieved on 30 March 2020].

O'Connell, A., Kousoulidou, M., Lonza, L. and Weindorf, W. (2019). Considerations on GHG emissions and energy balances of promising aviation biofuel pathways. Renewable and Sustainable Energy Reviews 101: 504–515. doi:10.1016/j.rser.2018.11.033.

Pechstein, J., Bullerdiek, N. and Kaltschmitt, M. (2020). A "book and Claim"—Approach to account for sustainable aviation fuels in the EU-ETS – Development of a basic concept. Energy Policy 136: 111014. doi:10.1016/j.enpol.2019.111014.

Peeters, P., Higham, J., Cohen, S., Eijgelaar, E. and Gössling, S. (2018). Desirable tourism transport futures. Journal of Sustainable Tourism, 1–16. doi:10.1080/09669582.2018.1477785.

Perera, A.I. (2009). CO_2 emissions in commercial aviation. Spain. Retrieved from http://www.divulgameteo.es/uploads/Emisiones-CO2-aviaci%C3%B3n.pdf [retrieved on 30 March 2020].

Rosado-Serrano, A., Paul, J. and Dikova, D. (2018). International franchising: A literature review and research agenda. Journal of Business Research 85: 238–257.

Rudall, P. and Bateman, R. (2002). Roles of synorganisation, zygomorphy and heterotopy in floral evolution: The gynostemium and labellum of orchids and other lilioid monocots. Biological Reviews of the Cambridge Philosophical Society 77(3).

Scheelhaase, J., Maertens, S., Grimme, W. and Jung, M. (2018). EU ETS versus CORSIA—A critical assessment of two approaches to limit air transport's CO_2 emissions by market-based measures. Journal of Air Transport Management 67: 55–62.

Schneider, L., Maosheng, D., Stavins, R., Kizzier, K., Broekhoff, D., Jotzo, F., Winkler, H., Lazaro, M., Howard, A. and Hood, C. (2019a). Double counting and the Paris Agreement rulebook. SCIENCE, [online] 366(6462): 180–183. https://science.sciencemag.org/content/sci/366/6462/180.full.pdf?ijkey=92joBj7Mq.3dI&keytype=ref&siteid=sci [retrieved on 30 March 2020].

Schneider, L., Michaelowa, A., Broekhoff, D., Espelage, A. and Siemons, A. (2019b). Lessons learned from the first round of applications by carbon-offsetting programs for eligibility under CORSIA. Oeko-Institut e.V. https://www.oeko.de/fileadmin/oekodoc/Lessons-learned-from-CORSIA-applications.pdf [retrieved on 30 March 2020].

Shi, X., Wang, K., Shen, Y., Sheng, Y. and Zhang, Y. (2020). A permit-trading scheme for facilitating energy transition: A case study of coal capacity control in China. Journal of Cleaner Production 256: 120472. doi:10.1016/j.jclepro.2020.120472.

Somerville, H. (2003). Transport energy and emissions: aviation. pp. 263–278. *In*: Hensher, D.A. and Button, K.J. (eds.). Handbooks in Transport 4: Handbook of Transport and the Environment. Elsevier.

Soria-Baledón, M. and Kosoy, N. (2018). "Problematizing" carbon emissions from international aviation and the role of alternative jet fuels in meeting ICAO's mid-century aspirational goals. Journal of Air Transport Management 71: 130–137. doi:10.1016/j.jairtraman.2018.06.001.

Staples, M.D., Malina, R., Suresh, P., Hileman, J.I. and Barrett, S.R.H. (2018). Aviation CO_2 emissions reductions from the use of alternative jet fuels. Energy Policy 114: 342–354. doi:10.1016/j.enpol.2017.12.007.

Verifavia. (2019). Which States are participating in CORSIA? https://www.verifavia.com/greenhouse-gas-verification/fq-which-states-are-participating-in-corsia-250.php [retrieved on 30 March 2020].

Wardle, D.A. (2003). Global sale of green air travel supported using biodiesel. Renewable and Sustainable Energy Reviews 7: 1–64.

Warnecke, C., Schneider, L., Day, T., La Hoz Theuer, S. and Fearnehough, H. (2019). Robust eligibility criteria essential for new global scheme to offset aviation emissions. Nature Climate Change 9: 218–221.

Winchester, N. (2019). A win-win solution to abate aviation CO_2 emissions. Journal of Air Transport Management 80: 101692.

World Business Council for Sustainable Development. Mobility-WBCSD. (2001). World Mobility at the End of the Twentieth Century and Its Sustainability. Published online: www.wbcsdmotability.org [retrieved on 30 March 2020].

Zhang, L., Butler, T. and Yang, B. (2020). Recent Trends, Opportunities and Challenges of Sustainable Aviation Fuel. Green Energy to Sustainability: Strategies for Global Industries.

Zhang, Y., Li, S. and Zhang, F. (2020). Does an emissions trading policy improve environmental efficiency? Evidence from China. Sustainability 12: 2165.

Zheng, S., Ge, Y.-E., Fu, X. and Jiang, C. (2019). Voluntary carbon offset and airline alliance. Transportation Research Part B: Methodological 123: 110–126.

CHAPTER 9

An Introduction to the Status of Tourism in Mexico

Camilo José Medina Ramírez

Introduction

In the literature, there are many definitions of tourism; each of these definitions is correct but incomplete at the same time.

According to United Nations World Tourism Organization (UNWTO), tourism is "the activity of people traveling and staying far from their usual setting for no longer than a year, but more than 24 hours for leisure, business, or other purposes" (UNWTO 2010). The UNWTO definition attempts to provide rules for excluding other traveling phenomena, such as migration, research expeditions, diplomatic trips, or military incursions. The UNWTO definition also tries to set two rules for traveling-related issues: exclusive association of traveling with leisure activities and a time frame for traveling.

Tourism could also be defined as the action of traveling to and from places outside one's home (Bauman 1998). According to this definition, people are tourists almost all the time. When someone travels away from home, that person becomes a tourist (Urry 2012). Tourism is not just an action; it is a way of thinking about and seeing the world at the end of a long, arduous journey (Franklin 2003). Tourism is more than traveling and leisure; it is a serious commercial model, with strong political and social implications (Smith 1998). This definition, provided by Thomas Cook, is the benchmark par excellence of tourism in the contemporary world.

The definitions by Cook and the UNWTO have been often selected as a theoretical frame, because they apply to the real world, and are easy to understand. These are also the only definitions that have correctly delimited the touristic phenomenon, as well as its implications on different fields of study.

Facultad de Ciencias, Universidad Nacional Autónoma de México (UNAM), Av. Universidad 3000, Circuito Exterior S/N Delegación Coyoacán, C.P. 04510. Ciudad Universitaria, CDMX, México; camilo@ciencias.unam.mx; medinaramirezcamilojose@gmail.com

Today, tourism is a social practice that is often a requirement for a person to be regarded as a "modern individual" in certain social groups, particularly by people of high social status. This encourages the formation and maintenance of a certain elite of traveling people. The farther and longer persons travel outside of their place of residence with leisure purposes, the higher their power for understanding the world.

Tourism is a field of knowledge that has received little attention from academic studies, despite being a phenomenon that attracts considerable investment and technological innovations, and that has great implications for information sciences, for communication, and for sociocultural studies.

Visiting and staying in a receiving community depends on time, which is a limiting factor for people going on a trip. Residing in an area, even if it is for less than a day, implies the use of basic resources on the part of the visitor (food, water, waste production, management, etc.). Resource consumption increases the longer the person resides in that area (Cordero 2009).

Tourist Attractions

The reason for people traveling to any point of interest has to be delimited. A 'tourist attraction' is defined as any metaphorical, conceptual, physical, or intangible thing that a visitor desires to visit or finds striking in a chosen destination (Pike and Page 2014).

A place or location with one or more tourist attractions is called a tourist destination. This place will cover the necessities of visitors and will absorb any damage caused voluntarily or involuntarily by visitors (Saraniemi and Kylänen 2011). Some examples of tourist attractions are forests, national parks, flora and fauna reserves, different ethnic communities, buildings and structures, historical and archeological sites, cultural and sports events, artistic manifestations, galleries and museums, souvenirs, botanical gardens and zoos, theme parks, viewpoints, food, festivals and concerts, places associated to feelings, etc. The reasons for wanting to experience those 'attractions' depend on the value each visitor deposits on them, and on their expectations and previous experiences (Beirman 2003).

The tourist market establishes the exchange of money as a requirement for visitors to have a temporary stay in the zone selected. The parts that make up this social interaction are the visitor, who does not live in the receiving community and uses their services for a short time, and the receiving community, who are the people who live in the site visited and who satisfy the needs and provide leisure activities for the visitor, in exchange for economic remuneration (Tassiopoulos 2011). This interaction also bears labor, economic, and sociocultural conflicts (due to the consumption of natural resources and the production of waste), and also changes in the behavior and cultural values of the receiving community.

A tourist attraction has two components: classification of tourism type (ecotourism, architectural tourism, health tourism, etc.), and the strategy of commercial exploitation of tourism. To date, mass tourism is the most popular strategy of commercial tourism, and it generates resource consumption and waste problems. Mass tourism is a business model based on the massive displacement of people (visitors). Mass tourism started as a commercial practice in several nations during the second half of the 20th century after the Second World War came to an end, when the sale of expectations and experiences began in different countries to attract tourists (Claver-Cortés et al. 2007).

Mass tourism is distinguished by an approach in which the visitor's entertainment and amusement needs are satisfied. This market strategy is designed for people who are interested in buying cheap holiday packages that include accommodation and transportation. The characteristics of the place do not matter as much as the price. The best examples of this kind of destination are sun and beach destinations, which are the most popular internationally (Rábago and Revah 2000). Mass tourism is grand scale tourism focused on supplying a massive demand (Zamorano 2002).

Mass tourism is characterized by low interaction of visitors with the receiving community, and by mobility reduced to the minimum. Mass tourism, or enclave tourism, is based on hotel complexes or big tourism centers that are designed to be alien to their social environment; they are similar to one another and have different surroundings. The services that they provide are set apart for visitors, excluding and marginalizing the local population, which causes the economic spill to benefit the receiving community only rarely or opportunistically (Manning 1996). Unfortunately, these touristic development projects have a harmful impact on the ecological and sociocultural aspects of the community (Ibañez 2007).

The classification of tourist attractions has been suggested as a way of knowing the characteristics, implications, and mitigations that the different types of tourism have on environmental and sociocultural levels. However, a clear classification for tourist attractions and types of tourism does not exist (Boullón 2006).

There is a hierarchy for tourist attractions proposed for Latin America, promoted by the Centro Interamericano de Capacitación Turística (CICATUR). This hierarchy is designed to work as a marketing tool, and is based on the geographical characteristics of tourist attractions; it examines visitors' perceptions to classify the tourist attraction in four levels of tourism impact (0–3): level three means that a tourist attraction is of interest to the international market and that is representative for all mankind, motivating tourist visits by itself; level two is an attraction with representative features of a country's culture, with or without additional aspects that may motivate the visitor; level one is an attraction of interest for certain types of visitors (mainly for the local tourist market), which is not interesting enough to create a market on its own; level zero is an attraction with insufficient merits to attract visitors, but that is still a part of the cultural heritage of humanity or can complement tourist attractions of a higher hierarchy level (Navarro 2015). Even if this tourist attraction classification attempts to be as holistic as possible, it is likely that a classification of tourist destinations that takes into account every relevant factor for completely understanding tourist attractions will never be found. This reminds us that tourist attractions have to be analyzed by looking at the consumer behavior of visitors.

Tourism in Mexico

Tourism is an activity that has been used in Mexico as an opportunity to increase the macro and microeconomy of the country. In Mexico, there are public policies that promote the diversity of destinations (Cárdenas Tabares 2006, Gobierno de Chile 2013, Pérez-Ramírez and Antolín-Espinosa 2016).

Mexico has occupied a notorious place on the economic and tourism landscape since 1982, although it has not reached the level of investment and tourism development of many first-world countries (e.g., Italy, France, Spain, the USA, etc.).

Since the late 1950s, many Mexican territories have suffered from technification based on the bylaw termed "*Ley General del Turismo*", which was ratified in 1961. A legal foundation was thus established for future regulations on the tourism sector. This bylaw has only undergone one modification, which occurred during the presidency of Vicente Fox Quezada, in the same period in which the program termed "*Plan Nacional de Desarrollo Turistico*" was established. This program set efficiency and productivity as priorities for the development and construction of hotel complexes and modern spaces for tourist reception; in other words, this plan promoted the enclave tourism business model (Pérez and Villa 2011).

Tourism has become fundamental to the Mexican economy. Tourism directly represents 8.5% of the country's gross domestic product (GDP). According to data shown by the *Secretaría de Turismo* (SECTUR), Mexican tourism represents around 10 million jobs (four million direct jobs, and six million indirect and induced jobs).

In 2015, a record of 32.1 million international tourists was established in Mexico, generating earnings of 15.5 billion USD. Ninety-one percent of the tourist exports consist of added value generated internally, which is superior to the average of the Organization for Economic Cooperation and Development's (OCDE) of 80 percent.

Cross-border or international tourists are the main sources of income for Mexican tourism because visitors spend more money and remain longer than national tourists, which represents a significant contribution to the country's economy. In 2015, international tourists accounted for only 21% of tourism but contributed with 84.4% of the income generated by tourism.

In Mexico, international tourism focuses 84.7% of the demand on five destinations, all of them sun and beach destinations. The most important destinations in the country are Cancún, Riviera Maya, and Los Cabos. The state of Quintana Roo alone represents almost half of all international arrivals for tourism or leisure purposes.

The number of international tourists hits its highest peak in 2016, with an increment of 50% over the number of visitors over the last decade, from 23 million to 35 million. During the last decade, the currency income from unknown tourists increased by 53%, going approximately from 12 to 20 billion dollars. These changes caused Mexico to be positioned in sixth place of the most visited countries in the world.

International visitors staying for longer periods cause larger resource expenses and sociocultural behavioral disturbances in the areas that they visit. On the other hand, national tourists represent the basis for the maintenance of the Mexican tourism sector, because they have easy access and means of transportation to different areas of the country. The frequency of visits of national tourists is much higher than that of foreigners, and there is a homogeneous distribution of national tourists through the territory.

Internal tourism represents an economic contribution to places that international visitors ignore. The income from internal tourism in 2017 was between 88 and

100 million pesos, which is enough to keep the tourist market afloat during non-holiday periods in a year.

According to data shown by tourist monitoring system DATATUR, by January 2019, the arrival of 3.5 million international tourists was registered, exceeding by 143,000 tourists the number of tourists in January 2018. The foreign exchange earnings per international traveler rose to 2,289 million USD, which represents an increase of 17.7% compared to January 2018. In 2019, the currency spent by Mexican travelers going abroad was 913 million USD, which was lower than the 102 million spent in January 2018, representing a decrease of 10.1 percent. In 2019, the departments of international tourists abroad decreased to 1,544,000 USD, which is 142,000 USD less than in January 2018, representing a reduction of 8.4 percent. Based on this information, it is estimated that in every region, except for northern Central Mexico, a weakening of the economic performance in the first quarter of 2018 was registered (Torruco and Levy 2019).

In terms of tourist attractions, Mexican tourism has to offer 187 archeological sites, 11,000 km of coastal territory, 34 world heritage sites, and 121 *Pueblos Mágicos* (Magical Towns). Today, Mexico is the seventh most visited country in the world with 41.5 million visitors in a year (Guerrero and Heald 2015). However, there is a disparity between the number of visitors and the income related to tourism (Mexico is in sixth place with 22,000,500 USD). Also, the tourism sector in Mexico has been reported to face various competitiveness and sustainability problems (UNWTO 2019).

As stated in the report *Estrategia Nacional de Turismo para el sexenio de 2019–2024*, written by the current secretary of tourism Miguel Torruco: "The social and regional benefits are far from macroeconomic results since they haven't permeated in every habitant, which makes touristic paradises coexist with margination hells" (Torruco 2019).

In summary, Mexico had a moderate economic activity in the past few years and had a growth of 2.6% in the projected GDP in 2016. Later on, it experienced five years of slow economic growth, low productivity, informal job market, and high inequality of income.

Top 18 tourist arrivals		In USD billion 2017	2018	Top 18 tourist arrivals		In USD billion 2017	2018
1	France	86.9	N.A.	1	United States	201.7	214.5
2	Spain	81.9	82.8	2	Spain	68.1	73.8
3	United States	76.9	N.A.	3	France	60.7	67.4
4	China	60.7	62.9	4	Thailand	56.9	63
5	Italy	58.3	62.1	5	United Kingdom	49	51.9
6	Turkey	37.6	45.8	6	Italy	44.2	49.3
7	Mexico	39.3	41.4	7	Australia	41.7	45
8	Germany	37.5	38.9	8	Germany	39.8	43
9	Thailand	35.5	38.3	9	Japan	34.1	41.1
10	United Kingdom	37.7	N.A.	10	China	38.6	40.4
11	Japan	28.7	31.2	11	Macao, China	35.6	40.2
12	Australia	29.5	30.8	12	Hong Kong, China	33.3	63.7
13	Greece	27.2	30.1	13	India	27.4	28.6
14	Hong Kong, China	27.9	29.3	14	Turkey	22.5	25.2
15	Malaysia	25.9	25.8	15	Austria	20.4	23
16	Russia	24.4	24.6	16	Mexico	21.3	22.5
17	Portugal	21.2	N.A.	17	Canada	20.3	21.9
18	Canada	20.9	21.1	18	United Arab Emirates	21	21.4

Figure 9.1. Comparison of countries with the highest number of tourism reception (left) and the countries with the highest income (right). Adapted from DATATUR (2017).

Sustainable Tourism

Due to consumer behavior, mass tourism causes local, regional, and international conflicts of interest. There is a generalized lack of consideration of the impact of tourism on the receiving community and on the environment. Conflicting consumer behaviors include the high demand for sophisticated services, overproduction of waste that alters biogeochemical cycles, biodiversity predation, and neoliberal practices. These behaviors degenerate the local culture, damage traditions, and slowly cause loss of cultural values, segregating low-income people, and resulting in regional debts (Rivas 1998, Rodríguez 2010, Pérez and Villa 2011, UNWTO 2018).

Sustainable tourism is an alternative to mass tourism and its many problems. The UNWTO defines sustainable tourism as "the tourism that takes into consideration its economic, social, and environmental impacts, current and future, tending to the visitor's needs, the industry, and the receiving community's environment" (WT and TC and APEC 1998). The term sustainable tourism came into use by the late 1980s, and has been firmly established in politics, development strategies, and tourism investigations ever since (Budeanu et al. 2016, Orgaz and Moral 2016).

Sustainable development means the conservation and regeneration of natural capital when needed. The sustainable tourism strategy, backed by the idea of sustainability, satisfies current tourism needs without compromising the capacity of future generations. This is an ambitious vision of tourism because it implies knowing all types of tourism and interacting with them. This commercial strategy not only takes into account economic benefits, but also tries to recognize a broader problem, seeks to make tourism more rational and conscious, and tries to make visitors responsible for their actions when they get in touch with nature and with the community. Many of these practices include equal distribution of income in the communities, open and supportive local companies, the promotion of culture, local

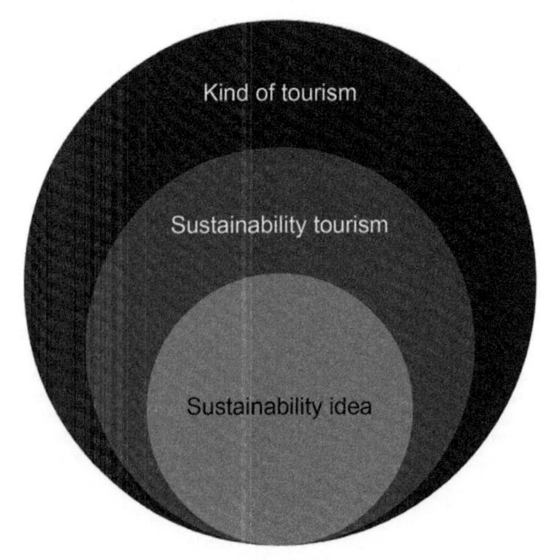

Figure 9.2. Origin, effect, and role of the sustainable tourism model.

tradition, and environment conservation (Gobierno Federal 2002). These practices aim for the improvement of the quality of life and have an active role in the labor aspects of the receiving community.

Since the beginning of president Vicente Fox's six-year term (2000–2006) until now, the best-positioned tourist destinations are non-metropolitan cities, sun and beach destinations are in a middle range, while the least visited ones are natural tourism destinations. The results show better performance indicators for urban developments, while the largest pressure is on the natural areas.

By the beginning of the 21st century, Mexico started to assess its tourism destinations internally using the report "*Agenda 21*" as a basis. This administrative process presents Sustainable Development Goals (SDG) indicators evaluating four topics: environment, socioeconomic environment, tourist arrival-departures, and urban development (Sui-Qui and Leng 2015).

In 2015, Mexico was one of the 193 Member States of the United Nations that approved the 2030 Agenda for Sustainable Development, in which 17 SDGs were established. As a response, and to improve sustainable participation in the tourism sector, the Economic Commission for Latin America and the Caribbean (ECLAC) launched a set of programs that year for interdisciplinary technical training and intergovernmental regional platforms to teach its member states (including Mexico) different ways to accomplish the Agenda's goal (Cepal 2015).

Since 2015, an editorial program for scientific divulgation has been promoted, as well as a program for environmental education, training, and communication. Currently, Mexico has 79 sustainable tourism committees distributed through its territory, from which private and public companies, actors of the tourism sector, academics, and citizens can get information on how tourism can contribute to the items proposed by the sustainability framework (Assembly 2002). The institution *Centro de Educación y Capitación para el Desarrollo Sustentable* (CECADESU), part of the *Secretaría de Medio Ambiente y Recursos* Naturales (SEMARNAT) under the coordination of the Federal Government, distributed printed materials over the last decade to promote environmentally-friendly practices for tourism enterprises. They also draw attention to the impacts generated by tourism and how they can be mitigated while promoting the efforts made by Protected Natural Areas (SEMARNAT 2017b).

In 2017, the year of sustainable tourism, the Travel and Tourism Competitiveness Report from the World Bank, pointed out that Mexico was in second place in natural resources, but in 116th out of 136 places in the sustainability category. That same year, an interview with Francisco Madrid Flores, the director of the Tourism Faculty at *Universidad Anahúac del Norte*, pointed out that "it is time to make mass tourism, which reigns in Mexico, sustainable" (Anahúac 2019).

In 2017, investments were done to support small and medium-sized enterprises, as a more direct action from the Federal Government to promote sustainable tourism in Mexico. Investments were aimed at the development of tourist real-estate development, green hotels, tourist marinas, clean beaches, and ecotourism, as examples of success that Mexico had in the past five years (SEMARNAT 2017a).

To increase sustainability efforts, Mexico counts with sustainable tourism evaluation measurements. The government bodies in charge of these evaluations are the *Comisión Nacional para el Conocimiento y Uso de la Biodiversidad* (CONABIO) and the *Comisión Nacional de Areas Naturales Protegidas* (CONANP). To date, Mexico has 91 ecotouristic projects distributed throughout the territory (supported by CONABIO and CONANP) with certificates that endorse them as sustainable tourist destinations (CEPAL 2015). One of the greatest limitations of some projects that receive these kinds of certificates is that they are regarded as "fake-ecotourism". Often, the only eco-friendly item they have is the tourist development itself, often based on a natural phenomenon, such as the migration of animals or exuberant vegetation; such sites have no methods of evaluation to prove that tourism is not degrading the area visited (Rodríguez 2010).

Despite a well-defined theoretical framework, sustainability in the tourism sector is still just a proposal. Mexico is still not 100% sustainable, and has a long way to go. The challenge is to sustain a place's economy based on a functional tourism model while avoiding the vices of enclave tourism. In other words, the problem resides in the way that a successful coexistence of factors and interests is carried out. A standard or a right way to accomplish sustainable tourism does not exist; however, identifying the trends in the visitor's consumer behavior, and the main existing problems, may help. This constitutes work that requires the participation of different actors, such as investors, investigators, activists, residents, etc.

The most important problem for sustainable tourism is to overcome methodological challenges. Sustainable tourism in Mexico has norms and laws that regulate its qualification, and government agencies that boost its development under value chain terms. Research results show that sustainable tourism exists in politics at a national level, but is not a practice that the industry undertakes. Nonetheless, tourists play an important role in development. If tourists demand eco-friendly and sustainable products, coercion will take place in the market and the actors will move forward with the paradigm of developing sustainable tourist areas (López et al. 2016).

The program *Pueblos Mágicos de México* is a big effort by the Mexican Federal Government in the past couple of decades, aiming to develop towns that wish to establish a sustainable tourism model. The program *Pueblos Mágicos de México* was created in 2001 under the initiative of the former secretary of tourism Leticia Navarro Ochoa; the program started with three locations designed to promote a new tourism vision. Leticia Navarro stated that the communities in localities that wanted to achieve economic self-sufficiency based on tourism would represent all the tourist potential Mexico has to offer in the shape of a wide variety of destinations and types of tourism, attracting national and international tourists.

Specifications that must be met for an area to be a part of the *Pueblos Mágicos* (Magical Towns) program are: having at least 200,000 inhabitants and being closer than a two-hour journey from a tourist attraction or a population defined as market hostess, such as Cancún, Los Cabos, Acapulco, etc. (Gross 2011, Ruiz Pelcastre 2014). The government body *Secretaría de Turismo* defines a Magical Town as "a locality that has defended its historical and natural patrimony through time and that is expressed through tangible and intangible symbols, legends, history, transcendental actions, and everyday expressions". In other words, "it magically shows its sociocultural manifestations". This means offering a touristic product using the "best of the best" that the Mexican identity has to offer. Nowadays, 121 localities are part of the program and flaunt the title of *Pueblo Mágico* (López and López 2017).

Consumer Behavior

The organizations that manage tourist destinations demand the best possible information for decision making on the different areas of work. Therefore, Information and Communication Technologies (ITCs) are essential for strategic processes (López and Cendra 2018). ITCs will provide access to information about geolocalized activity, opinions on social media, flights and hotel prices, etc. Such information should be taken into account for decision making (Quesada et al. 2018).

How tourists perceive and use ITCs in tourist destinations with different characteristics constitutes a gap in research. Understanding how each type of destination attracts a certain market segment is necessary because the adoption and usage of technology can vary and affect the local tourist management (Femenia-Serra et al. 2018).

ITCs are a structural piece for tourist attractions, for enterprises, for destination management, and for consumer behavior (Benckendorff et al. 2014). Web 2.0 is the most important marketing tool. Destination marketing organizations (DMOs) use web 2.0 to reach a large number of people around the world and provide them with content and information about a location (Bernkopf and Nixon 2019). Internet is the main source of information for traveling, web apps, and the web 2.0. The internet provides access to information at every stage of a trip. The new era of technological innovation in tourism has diversified travel websites and web apps, has committed society, and has slowed down the decision-making process. Before, during, and after traveling, web 2.0 provides a holistic and dynamic approach (Yi et al. 2018).

The web 2.0 has also caused a change in customer profiles. The visitor's consumer behavior that dominates in a given touristic area can be identified based on the images of a tourist destination, particularly the ones created by the consumers (also known as tourists 2.0) or "ad prosumers". These consumers are people who use the web 2.0 (social media) to recommend, promote, and consume a touristic product (Zhang et al. 2008). As the competitiveness between tourist destinations increases, tourists 2.0 tend to search for greater enjoyment with the least expenses. Therefore, the current era is based on technology and the use is given to the internet and mobile apps at every stage of the travel (Vigolo 2017).

As mentioned before, tourist attractions constitute an important component of the visitor's destination choice. Increasing the number of visitors to a tourist destination can be achieved by improving the tourist experience. Access to the most demanded tourist points of interest requires visual recognition techniques that use the information of a location (e.g., knowledge transfer, big data, the information generated by users, and social media analysis) (Quesada et al. 2018).

What makes tourism 2.0 attractive for marketing is the ability to impact the decision-making process of the consumer (Hudson and Thal 2013). The role of social media stands out within ITCs because ITCs can change the spectator's image of a tourist destination. The exchange of experiences and memories is shared on different platforms, such as Facebook, Instagram, Snapchat, etc. (Tiongson 2015). This kind of content is known as Electronic-Word-of-Mouth. For tourism 2.0 studies, "the use of social media amongst the main DMOs is still a great experimental measurement and the strategies vary significantly" (Hays et al. 2013). Because of its global reach, the speed at which it travels, its ease of use, its anonymity, and the absence of direct pressure, User Generated Content is more influential than DMOs marketing efforts (Lange-Faria and Elliot 2012).

The best way of convincing a user to purchase a touristic product in this context is by allowing consumers to promote their visit. This social media phenomenon works as a way of comparing expectations and stereotypes. Showing the possible experiences of an ongoing traveler to the potential consumer can trespass geographical barriers. When the trip comes to an end, social media become important databases, because any user can use them to give opinions and provide evidence of the tourist attractions that exist in a place of interest. This post-trip effect has caught the eye of many social media-based enterprises. Nowadays, the sale and purchase of metadata to companies (including tourism-based companies) often takes place to avoid tedious and expensive sampling processes to generate a faster buyer profile (Amboage 2011).

In web 2.0, the update of customer segment information is constant, a disadvantage of which is that it can also attract criticism and fake or biased recommendations to a particular touristic product. Regardless, people will keep using web 2.0 as a reference for tourism-related subjects, and this tendency is expected to grow continuously, revolving around the image of a given tourist destination.

The image of a tourist destination can be analyzed into two components: cognitive and affective. The cognitive component studies beliefs or knowledge about the characteristics, actions, or objects in a tourist destination; on the other hand, the affective component shows the feelings that a person has towards a tourist destination. There is a significant influence of the cognitive component on the

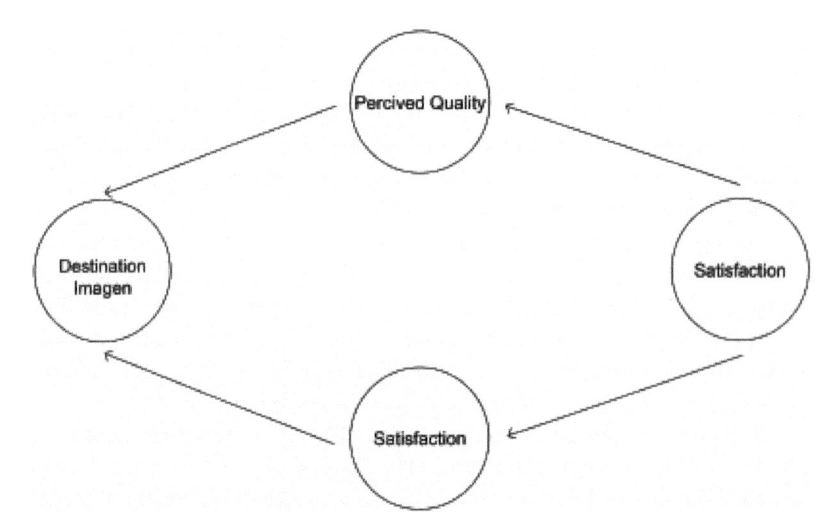

Figure 9.3. Relationships between destination image, perceived quality, emotional place attachment, tourist satisfaction, and post-visiting behavior intention. Marketing review. Adapted from Li et al. 2015.

affective component (Baloglu and Brinberg 1997, Kim and Yoon 2003, Beerli and Martín 2004, Pike and Ryan 2004, San Martín and del Bosque 2008, Huang et al. 2010, Day et al. 2012, Zeng et al. 2015).

The process of manual categorization to identify common patterns through the analysis of tourist attractions is done inductively (Bosangit and Mena 2009) because categories are obtained once the information has been collected through data based on the examination of patterns and recurrences of tourist destination images (Chaves 2005).

According to the approach of new methodological proposals by Huang et al. (2014), the analysis of the images generated by tourists 2.0 has to meet a series of demands:

- Image identification through geolocation: The areas selected can be located on a map, which allows the comparison with maps from other official or social media servers. The purpose is to reduce the ambiguities of the images and to verify that their content corresponds to the same location.
- Confirmation of the visitor's profile: The selected social media can verify the identity of the user who is sharing images and links of a location. This allows us to establish if they are visitors or local people at a tourist location of interest, in order to avoid mixing visitors with local people. Visitors are the group of interest and, if desired, different categorizations can be done before performing a study.
- Examination of pattern selection: It allows discrimination between what is considered attractive and tourist behavior, and to associate with a tourism framework (Huang et al. 2014). For the present study, "Eye-Catcher" analysis is selected. This method excludes from the image description "selfies", intentional advertisements, elements that do not correspond to recreational activities, and natural or man-made objects portrayed by the visitors (Paül i Agusti 2019).

Research regarding social media perception related to tourism has never been conducted, which is why an analysis of the tourist trends preferred by visitors and their effects on sustainability in Mexico is elusive.

This chapter attempts to clarify the trends of tourist attractions, and the visitor's consumer behavior in Mexico, associated with negative effects on tourist sustainability at locations where the population has chosen to develop a sustainable tourism approach.

Methodology

The present study is prospective, non-experimental, and *ex-post-facto*, which means that there is no variable manipulation, the variables are pre-existing and cannot be tampered with.

The visual elements, Image Hits (IH), that correspond to tourist attractions and that are associated with negative effects for achieving sustainable tourism in Mexico are the independent variable. The grouping of several types of tourism and the analysis of patterns and trends is the dependent variable. To meet the objective

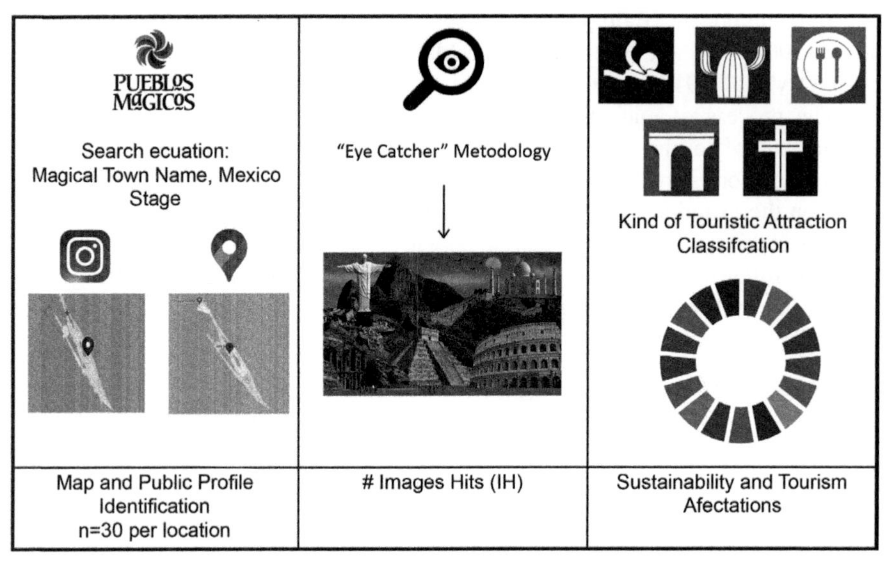

Figure 9.4. Manual categorization of tourist attractions. Elements to be taken into account are: the sampling points, comparison between web 2.0 sources, selection of analysis type for extracting visual elements, classification of tourist attraction types, and sustainability effects.

of analyzing the visual representations generated by tourists 2.0 in Mexico, the communities of the locations that belong to the program of *Pueblos Mágicos de México* were chosen as the object of study because these communities agreed to turn their towns into tourist centers in line with a sustainable tourism strategy.

The following methodological proposal aims to find classification trends of tourist attractions and to identify the relationship with the predominant effects on Mexico's sustainability.

For social media, Instagram and images from Google Maps were chosen for this investigation because they are open access, and because of their popularity. The following image analysis requirements were set:

- Images must have a location identification system that can be visualized on each app, allowing image comparison. On Instagram search results, there is a map on the top of the image feed associated with a location. This Instagram map can be compared with the one on Google Maps for location confirmation, and comparison of images associated with both search engines.
- The visitor profile must be identified in both social media, which can only happen when a user sets the profile as public; this allows the separation between visitors and residents. There was no discrimination between local and foreign tourists.
- The eye-catcher methodology was included to classify the visual elements or IH that are also tourist attractions.

Sampling was direct, prospective, and crossed. Locations classified as Magical Towns (121 localities) were chosen because of the goals of the *Pueblos Mágicos de México* program described above. These goals are promoted by the town's local community, who seek a tourism-based economy using their tourist attractions as an emblem, and as an exclusive and prestigious brand.

Sampling took place from 8 January to 8 February 2020. For each Magical Town, 30 images were selected in order of appearance, making sure they met the three requirements described above. The content of each image was analyzed. The total number of images analyzed was 4,830.

The number of appearances of visual elements (objects or activities), understood as tourist attractions, was counted and visual elements were grouped as IH. Finally, IH were classified into five types of tourism, which were designated after the identification of tourist attractions, in order to avoid possible biases towards one of the tourist attractions.

IH with a low number of appearances were discarded. To carry out the selection at each location, tourist attractions with the highest number of HI were selected and this number was divided by two. Only tourist attractions within the upper IH range were included. Then, the IH of tourist attractions at every location were added together and plotted as an accumulation curve for each social media selected.

Results

The effects that tourist attractions have on sustainability in Mexico can be reviewed in more depth and individually in Appendix 1. Here, a graphic summary of the observations will be described using data collected from the IH of each social media. The objective is to point out trends in tourist attractions portrayed by tourists 2.0.

Figure 9.5 shows the five-group classification of tourism types and the frequencies of IH that set trends in Mexico in each social media.

For the classification of tourism types, a breakdown of Figure 9.5 was performed. For each tourism type, different numbers of tourist attractions were found. Tourist attractions were classified by color and associated with an icon for quick identification. Lastly, tourist attractions were associated with an effect that could interfere with the development of a tourism strategy based on sustainability. These affections were found in the literature in studies related to tourism problems, and in news reporting tourism problems in Mexico.

The trends reported in Appendix 1 are summarized as follows:

- Buildings tourism type (in pink) consisted of six tourist attractions that set the trend for the formation of the group. The problems interfering with tourism sustainability in this group are light pollution, disturbance of circadian rhythms, alterations of wildlife behavior, cultural heritage damage, solid waste generation, fireworks pollution, heat island effect, waste of water, insect infestation, chemical exposure to repellents, damage to natural areas, and deforestation.
- Water activities tourism type (in blue) consisted of five tourist attractions that set the trend for the formation of the group. The problems interfering with tourism sustainability in this group are sunscreen chemical pollution and marine fauna

perturbation, sediment carryover by boats, solid waste pollution, illegal fishing, water waste, and contamination with microplastics.

- Souvenirs tourism type (in orange) consisted of four tourist attractions that set the trend for the formation of the group. The problems interfering with tourism sustainability in this group are water waste, economic inequality, intellectual property lawsuits, racism against Mexican indigenous groups, and cultural damage.
- Celebrations and spiritual needs tourism type (in purple) consisted of three tourist attractions that set the trend for the formation of the group. The problems interfering with tourism sustainability in this group are solid waste generation, intellectual property lawsuits, economic inequality, and racism against Mexican indigenous groups.
- Environment tourism type (in green) consisted of six tourist attractions that set the trend for the formation of the group. The problems interfering with tourism sustainability in this group are forest fires, solid waste pollution, fake-ecotourism, natural and anthropic erosion, water waste, insect infestation, damage to natural areas, animal abuse, lack of resources, deforestation, and environmental pollution.

Figure 9.5 shows tourist attraction trends classified into tourism types by search criteria in the two social media:

On Google Maps, IH of the buildings' tourism type were over 50% of the total number of IH. Using this search engine, a higher tendency towards the buildings' tourist attraction type was observed. IH associated with the environment tourist attraction type were the second most frequent ones, with a difference of approximately 30% the number of IH of the buildings' tourist attraction type. On this search engine, there were no IH associated with souvenirs.

On Instagram, IH showed a visible tendency towards the environment tourist attraction type. The second most frequent IH was the buildings tourist attraction type, with a difference of over 50% the number of IH of the environment tourist attraction type. None of the five tourist attraction types were absent in this social media, but

Visual elements category IH	Instagram IH	Google Maps IH	Visual elements category IH and percentage (%)
Buildings	696 (13.8%)	2111 (42%)	2807 (55.9%)
Aquatic activities	97 (1.9%)	128 (2.5%)	225 (4.4%)
Souvenirs	70 (1.3%)	0 (0%)	70 (1.3%)
[illegible]	72 (1.4)	102 (2%)	174 (3.4%)
Environmental	1218 (24.2%)	527 (10.5%)	1745 (34.7%)
Social Networks IH	2153 (42.8%)	2868 (57.1%)	5021 (100%)

Figure 9.5. Image Hits (IH) found in each social media per tourism type (percentage) to point out the contribution of tourist attractions to the preferred tourist trends portrayed by tourists 2.0 visiting Mexico. Author's creation.

IH associated with tourist attraction types souvenirs and celebrations, and spiritual needs were the least frequent.

Regardless of the social media, IH associated with the buildings tourist attraction type led the list, with a thousand units IH over the environment tourist attraction type.

On the other hand, accumulation graphs of IH of tourist attractions show that the tourist attractions that form the highest peak of the curve are the ones with the highest tendency to be exhibited by tourists 2.0.

Figure 9.6 shows that the tourist attractions that lead the curve of Instagram IH are (from the highest to the lowest number of IH): vegetation with 13.8% of IH (the highest percentage of IH was from Instagram), architecture with 10.5% of IH, street tourist attractions with 10% of IH, and rural tourist attractions (farms, plantations, fields, etc.) with 9.5% of IH. These numbers show that on the Instagram tourist attractions curve, environment tourist attractions predominate with 23.3% contribution to the total.

Figure 9.7 shows that the tourist attractions that lead the curve of Google Maps IH are (from the highest to the lowest number of IH): architecture with 19.7% of IH (the highest percentage of IH was from Google Maps), churches with 13.2% of IH, vegetation with 6.7% of IH, and street tourist attractions with 6.7% of IH. These numbers show that on the Google Maps tourist attractions curve, the buildings' tourist attractions predominate with 39.6% contribution to the total.

According to the information of Figures 9.6 and 9.7, there is only a small difference in percentages and the number of IH between the tourist attraction types that are not part of the peak of the curve. This does not mean that IH outside of the peak are not important in terms of the effects that tourism has on sustainability in

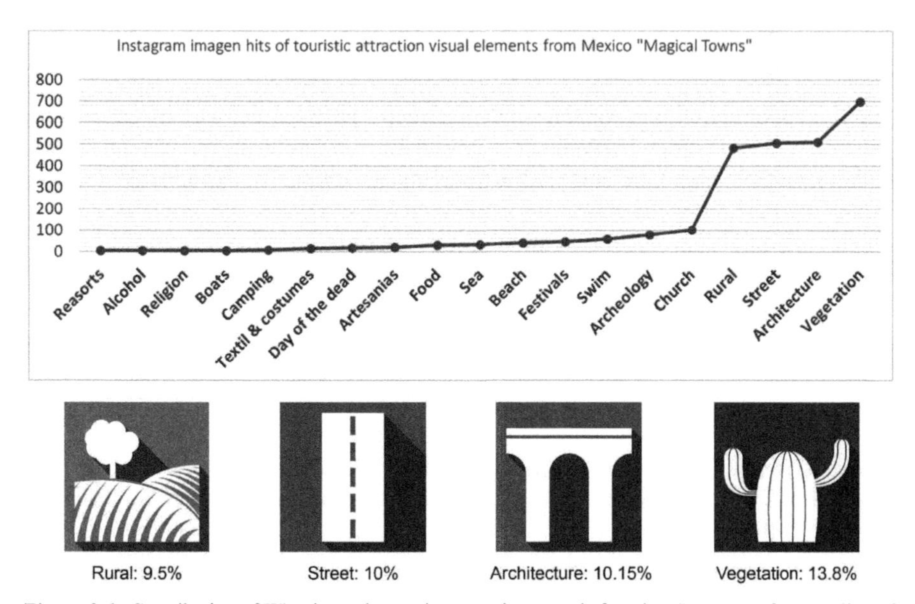

Figure 9.6. Contribution of IH units to the tourist attractions trends found on Instagram for sampling of locations corresponding to the program *Pueblos Mágicos de México*. Author's creation.

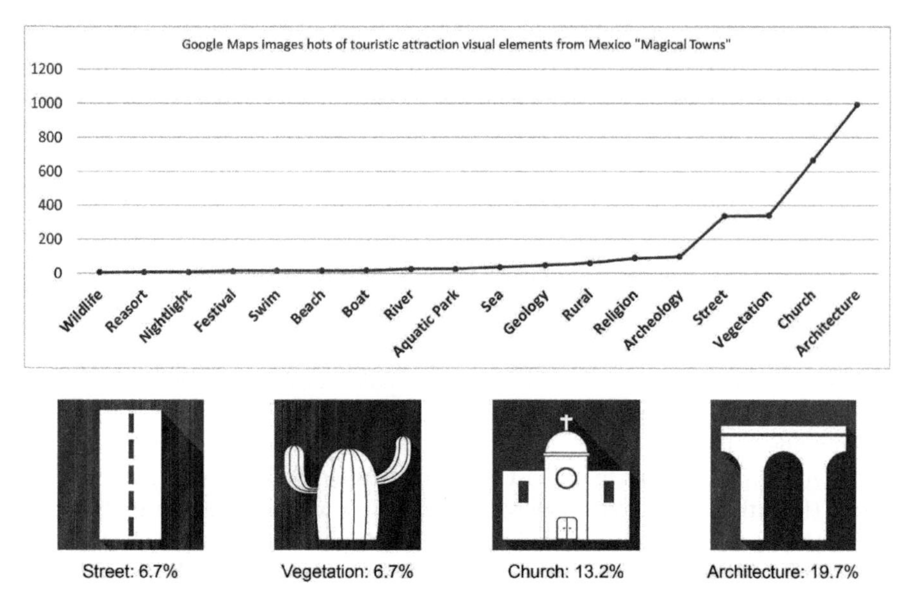

Figure 9.7. Contribution of IH units to the tourist attractions trends found on Google Maps for sampling of locations corresponding to the program *Pueblos Mágicos de México.* Author's creation.

Mexico or on the Magical Towns; this only means that the corresponding tourist attractions are the least favorite to be exhibited by tourists 2.0. The conclusion is that tourist attractions that involve a higher negative effect on the development of sustainable tourism in Mexico are the ones at the highest position on the table.

Discussion

The methodology chosen for this study indirectly portrays the way of associating habits and tourist attraction preferences in web 2.0 with problems already interfering with the development of sustainable tourism. Traditionally, sustainable tourism studies are performed the other way around; they first identify the problem and then a series of complex analyses are carried out to determine which of the tourists' habits originate them. This approach is more direct, but the possible answers to mitigate the negative effects of these habits are found more slowly. The association of tourist habits based on tourist destination images is an indirect, but holistic way to identify different tourist habits in a region. The application of this method allows faster identification of the main tourist interests of a location so that we can then focus on the implications that tourist attractions have on local sustainability.

A disadvantage of this methodology is that it can overlap tourist attractions and types of tourism. For example, the church tourist attraction type can be classified into two different types of tourism: buildings' tourist attraction type (because the architectural richness may be striking for tourists) or the religious tourist attraction type (because churches are a spiritual place of worship). In this study, the conflict was resolved by considering the outer part of the attraction as a building tourist attraction type and the inner part of it as a religious attraction. This strategy should be considered to avoid double counting a particular IH. Caution is advised when dealing with the contribution that a given tourist attraction may have on different kinds of tourism types, and with the sustainability problems associated with them.

In this study, the affective component of tourist attractions was set aside in the visual representation of the tourist destinations; we thought that if a trend was found on any kind of tourism, the affective component would have already acted and would already be present on the trends found, independent of the reasons underlying these trends.

There are not neutral times on the tourism periods, just a low or high influx of tourists. However, the tourist visits during vacation time cause the replacement of tourist destination-related images. Our sampling was performed at a time of low tourist influx so that we could learn about their opinions on new trends of tourist attractions.

Summer holidays are often associated with sun and beach destinations, while winter holidays are associated with more spiritual holidays. On the other hand, the social media selected (i.e., Instagram and Google Maps) save images associated with a location on different dates (some images were older than one year). In social media, image replacement is performed according to popularity states; however, it is unknown which particular criteria Instagram or Google Maps use to choose images to be posted as representatives of a location.

The eye-catcher type method of data analysis used in this study is a good system for classifying tourist attractions, but it presents the following inconveniences:

- It prevents the identification of tourist attractions: Tourists 2.0 have the habit of showing on social media that they have traveled to a certain place or to a tourist attraction without portraying any activities or tourist attractions in particular. This could cause an important bias, but in terms of the objective of this study, this was not considered a problem.
- There could be bias because there is only one person performing image selection. Having more than one person performing image selection would not remove the bias entirely. Thus, there should be a way of standardizing image classification trends. This methodological shortcoming should be taken into consideration for further data analysis.

Even if the eye-catcher method has some deficiencies, it is quite useful for the recognition of tourist attractions. An automatic alternative would be to perform this same study using visual recognition software and object-based image analysis to locate tourist attractions on a geographic information system (GIS). If these tools were used, a technological approach would be required to coordinate their use.

Our methodological approach could predict the next tourist attraction trend to be portrayed by tourists 2.0 in Mexico. According to our analysis (Figure 9.5), destinations that are likely to be portrayed by tourists 2.0 are buildings tourist attraction types. This means that urban areas are better positioned than other types of areas. The enclave model is still rooted in Mexican tourism and the efforts made to modify this model have not been successful. In 2017, Mexico was the second country with the most natural resources, but the 116th country out of 136 on sustainability (Anahúac 2019). The evidence presented here suggests that this situation has not changed. The worst part is still being received by the environment tourist attractions type, even if it is the second favorite destination category to be visited and portrayed by tourists 2.0.

For the past 20 years, Mexico has presented a stagnation in the tourist attractions model following the trends of attention and relevance of tourist destinations of the 2000–2006 period (Sui-Qui and Leng 2015). The *Pueblos Mágicos de México* program aims for their destinations to be sustainable. These communities are an example of a successful sustainable tourism-based economy. It would be important to carry out a study on how the holiday and non-holiday periods affect the local economy and the regional stability of natural resources. This would allow us to understand how, and which, Magical Towns are the most vulnerable in terms of sustainability and tourist attractions. A tourism marketing study could also be carried out to analyze the impact of tourism and perceptions from social media to improve their regional business model (Paül i Agusti 2019).

The different accumulation curves of Instagram and Google Maps are consistent with the most represented types of tourism and tourist attractions, specifically the buildings and environment types. The differences between them could be due to the particular characteristics of each social media, the purpose they are intended for, and the use given to them. Instagram is a network in which the users show the areas that they travel to, and occasionally the activities they perform, to a community of friends and followers. On the other hand, Google Maps is more focused on the display of locations and key elements as references to guide people through a particular area. Non-parametric statistical tests should be performed to validate the trends in the preferences of tourism types in Mexico that we found, independently of the social media used.

In our study, the elements that threaten the development of sustainable tourism are shared between the buildings and environment tourism types, regardless of social media. These problems are solid waste generation, water waste, insect infestation, alterations of wildlife behavior, chemical exposure to repellents, damage to natural areas, and deforestation. These problems must be solved urgently to achieve sustainable tourism in Mexico.

Conclusions

The limitations of this study are clear. One person alone collecting the tourist attraction IH could bias the investigation. The development of technological tourism management is necessary for the coordination of the geographical recognition of tourist attractions.

An image analysis system and a social network could be useful for a precise analysis of tourist attractions and of their implications in the consumer behavior of the tourists 2.0.

A geographical reference element would allow us to locate the most vulnerable areas affected by tourist practices in Mexico. It is necessary to develop a geographical information system using tourist attraction trends, types of tourism, and the negative effects that prevent the development of sustainable tourism.

To validate the trends of tourism types, it may be necessary to apply statistical tests. I suggest the Mann-Whitney test to determine if the differences observed are statistically significant.

I recommend a study on the role of the SDGs in the values system of the tourist actors in Mexico, addressing which actors are the most interested in sustainable development and why. With this information, we could identify the motivations of the visitors and of the receiving community, and also their conflicts with the SDGs.

I recommend approaching the receiving communities and/or the tourism investors of enclave tourism models to promote changes aimed at sustainable tourism models. These changes could directly influence the consumer behavior of travelers for the sake of the future tourism model in Mexico.

Appendix: Tourist Attraction Image Hits Table

Tourist attractions	Instagram IH	Google Maps IH	Sustainability challenges
Nightlight	-	8	Light pollution, disturbance of circadian rhythms, and alterations of wildlife behavior (Giraldo 2008, Chepesiuk 2009, Yorzinski et al. 2015, Bobkowska et al. 2016)
Archeology	79	99	Cultural heritage damage (Stovel 1998)
Church	102	667	Solid Waste generation, fireworks pollution (El Universal 2018)
Main Street	504	337	Heat island effect, solid waste generation, (Taha 1997, El Universal 2018)
Architecture	510	993	Heat island effect (Taha 1997)

Appendix Table Contd. ...

...Appendix Table Contd.

Tourist attractions	Instagram IH	Google Maps IH	Sustainability challenges
 Resort	5	7	Solid waste generation, water waste, insect infestation, chemical exposure to repellents, damage to natural areas, and deforestation (Peterson and Coats 2001, Kirk 2010, Vázquez 2019)
 Swimming	58	15	Sunscreen chemical pollution and marine fauna perturbation (Johnson et al. 2011)
 Navigation	6	18	Sediment carryover by boats (Warnken et al. 2004)
 Sea	33	37	Illegal fishing, solid waste pollution (Sumaila et al. 2006)
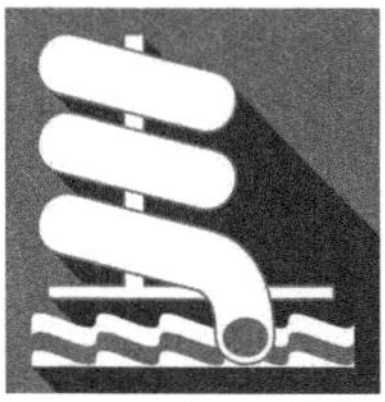 Aquatic Park	-	26	Water waste (Castellanos 2017)

Appendix Table Contd. ...

...Appendix Table Contd.

Tourist attractions	Instagram IH	Google Maps IH	Sustainability challenges
River	-	25	Microplastic contamination, sunscreen pollution, solid waste pollution (Lahens et al. 2018)
Alcoholic Drinks	5	-	Water waste, economic inequality (Excélsior 2018, Hegde et al. 2018)
Textiles	15	-	Intellectual property lawsuits, economic inequality, racism against Mexican indigenous groups, cultural damage (Johnston 2013, Arellano 2019)
Crafts	20	-	Intellectual property lawsuits, economic inequality, racism against Mexican indigenous groups, cultural damage. (Heredia and Francisco 2013, Johnston 2013)
Food	30	-	Intellectual property lawsuits, economic inequality (Johnston 2013)

Appendix Table Contd. ...

...Appendix Table Contd.

Tourist attractions	Instagram IH	Google Maps IH	Sustainability challenges
 Religion	6	89	Solid waste generation, intellectual property lawsuits (Johnston 2013, El Universal 2018)
 Day of the dead (DIA DE MUERTOS)	19	-	Solid waste generation, intellectual property lawsuits (Johnston 2013, El Universal 2018)
 Festivals	47	13	Solid waste generation, intellectual property lawsuits (Johnston 2013, El Universal 2018)
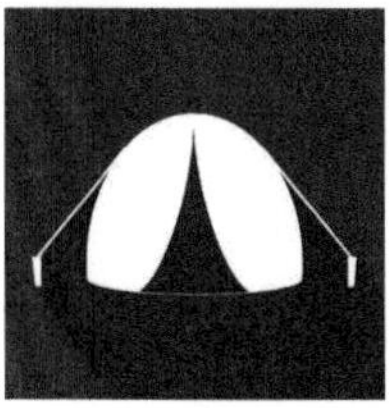 Camping	7	-	Forest fires (CONAFOR 2019)
 Rural	482	61	Fake-ecotourism (Rodríguez,2010)

Appendix Table Contd. ...

...Appendix Table Contd.

Tourist attractions	Instagram IH	Google Maps IH	Sustainability challenges
Geological Formations	-	48	Natural and anthropic erosion (Brilha 2018)
Beach	40	15	Solid waste generation, waste of water, insect infestation, damage to natural areas (Santos et al. 2005)
ZOO Wildlife	-	6	Animal abuse, lack of resources (Lira 2017)
Vegetation	693	341	Deforestation, environmental pollution, fake-ecotourism, alterations of wildlife behavior (Rodríguez 2010, Vázquez 2019)

References

Amboage, E.S. (2011). El turismo 2.0: un nuevo modelo de promoción turística. Redmarka: revista académica de marketing aplicado (6): 33–57.

Anahúac, G. (Red de Universidades Anahuac) Turismo sostenible, el Futuro de México, (2019). https://www.anahuac.mx/generacion-anahuac/turismo-sostenible-el-futuro-de-mexico.

Arellano, S. Senado demandará a Carolina Herrera por plagio de diseños indígenas. (Milenio diario). (2019). https://www.milenio.com/politica/senadores-alistan-demanda-carolina-herrera-plagio-disenos-indigenas.

Assembly, G. (2002). Resolution adopted by the General Assembly. Agenda 21: 7.

Bauman, Z. (1998). Globalization: The Human Consequences. Columbia University Press.

Beirman, D. (2003). Restoring tourism destinations in crisis: A strategic marketing approach. CAUTHE 2003: Riding The Wave of Tourism and Hospitality Research 1146.

Benckendorff, P.J., Sheldon, P.J. and Fesenmaier, D.R. (2014). The digital tourism landscape. Tourism Information Technology (Ed. 2): 22–52.

Bernkopf, D. and Nixon, L. (2019). The impact of visual social media on the projected image of a destination: the case of Mexico City on Instagram. pp. 145–157. In Information and Communication Technologies in Tourism. Springer, Cham.
Bosangit, C. and Mena, M. (2009). Meanings, motivations and behaviour of filipinoamerican first-time visitors of the philippines: A content analysis of travel blogs. In International Conference on Tourist Experiences, Lancaster, UK.
Boullón Roberto, C. (2006). Planificación del espacio turístico. Editorial Trillas, Cuarta Edición México DF.
Bobkowska, K., Janowski, A., Jasinska, K., Kowal, P. and Przyborski, M. 2016. Light pollution in the context of threats to the wildlife corridors. International Multidisciplinary Scientific GeoConference: SGEM: Surveying Geology and Mining Ecology Management 3: 665–670.
Brilha, J. 2018. Geoheritage: inventories and evaluation. In Geoheritage. Elsevier: 69–85.
Budeanu, A., Miller, G., Moscardo, G. and Ooi, C.S. (2016). Sustainable tourism, progress, challenges and opportunities: an introduction.
Cárdenas Tabares, F. (2006). Proyectos turísticos: localización e inversión. Trillas.
Castellanos, E. ¿Es sostenible el gasto de agua en los parques temáticos de ocio? (iagua, CETA2020). (2017). https://www.iagua.es/blogs/enrique-castellanos-rodrigo/es-sostenible-gasto-agua-parques-tematicos-ocio.
Cepal. (2015) CEPAL Agenda 2030 para el desarrollo sostenible. (UNWTO), 2015. https://www.biodiversidad.gob.mx/usos/turismo/.
Chaves, C.R. (2005). La categorización un aspecto crucial en la investigación cualitativa. Revista de investigaciones Cesmag 11(11): 113–118.
Chepesiuk, R. 2009. Missing the dark: health effects of light pollution.
Claver-Cortés, E., Molina-Azorı, J.F. and Pereira-Moliner, J. (2007). Competitiveness in mass tourism. Annals of Tourism Research 34(3): 727–745.
CONAFOR. Arranca CONAFOR operativo contra incendios forestales en periodo vacacional (Gobierno de México, comunicado, prense). (2019). https://www.gob.mx/conafor/prensa/arranca-conafor-operativo-contra-incendios-forestales-en-periodo-vacacional.
Cordero, J.C.M. (2009). Comunidad receptora: Elemento esencial en la gestión turística. Gestión turística (11): 101–111.
Day, J., Cai, L. and Murphy, L. (2012). Impact of tourism marketing on destination image: Industry perspectives. Tourism Analysis 17(3): 273–284.
DATATUR. (2017). Principales destinos turisticos en el mundo, Secretaria de Gobierno, 23/4/2019. https://www.datatur.sectur.gob.mx/SitePages/RankingOMT.aspx.
El Universal. Fiestas de fin de año generan 30% más de basura (Notimex, Ciencias de salud). (2018). https://www.eluniversal.com.mx/ciencia-y-salud/fiestas-de-fin-de-ano-generan-30-mas-de-basura.
Excélsior. Boicot contra Corona en defensa del agua. (Excélsior, Nacional). (2018). https://www.excelsior.com.mx/nacional/2018/02/26/1222827.
Femenia-Serra, F., García Hernández, M., Tuero, V. and Perles Ribes, J.F. (2018). Profiling tourists and their ICTs perception and use across Spanish destinations.
Franklin, A. (2003). Tourism: an introduction. Sage.
Giraldo, N. (2008). Impacto ambiental de los sistemas de iluminación. Contaminación lumínica. Empresas Públicas de Medellín https://www.grupo- epm.com/site/Portals/1/biblioteca_epm_virtual/tesis/impacto_ambiental_de_los_sistemas_de_iluminacion_luminica.pdf.
Gobierno de chile. (2013). (Gobierno de Chile, Servicio Nacional de Turismo). Propuestas metodológicas para la jerarquización, categorización y tipificación de Atractivos Turísticos de SERNATUR. (Subdirección de Estudios Unidad de Territorio y Medio Ambiente), 2013. https://biblioteca.sernatur.cl/documentos/680.983%20S491p.2013.pdf.
Gobierno Federal. (2002). Programa de Turismo Sustentable en México (Secretaria de Turismo) http://www.sectur.gob.mx/pdf/planeacion_estrategica/PTSM.pdf.
Gross, T. (2011). Divided over Tourism: Zapotec Responses to Mexico's 'Magical Villages Program'. Anthropological Notebooks 17(3): 51–71.
Guerrero, R. and Heald, J. (2015). El programa de Pueblos Mágicos ¿Contribución o limitación? La experiencia turística de Dolores Hidalgo. Revista de Arquitectura, Urbanismo y Territorios, 408–

425. https://www.hosteltur.com/lat/130054_la-mitad-de-las-divisas-por-turismo-se-concentran-en-10-paises.html.

Hays, S., Page, S.J. and Buhalis, D. (2013). Social media as a destination marketing tool: its use by national tourism organisations. Current Issues in Tourism 16(3): 211–239.

Hegde, S., Lodge, J.S. and Trabold, T.A. 2018. Characteristics of food processing wastes and their use in sustainable alcohol production. Renewable and Sustainable Energy Reviews 81: 510–523.

Heredia, S. and Francisco, J. 2013. Las artesanías en México situación actual y retos (338.6425 A7).

Huang, J.Z., Li, M. and Cai, L.A. (2010). A model of community-based festival image. International Journal of Hospitality Management 29(2): 254–260.

Huang, S., Van der Veen, R. and Zhang, G. (2014). New era of China tourism research.

Hudson, S. and Thal, K. (2013). The impact of social media on the consumer decision process: Implications for tourism marketing. Journal of Travel and Tourism Marketing 30(1-2): 156–160.

Ibáñez, R. (2007). Turismo alternativo, gestión y desarrollo local: El caso de Cabo Pulmo, BCS Tesis de maestría. uabcs. La Paz, México.

Johnson, A.C., Bowes, M.J., Crossley, A., Jarvie, H.P., Jurkschat, K., Jürgens, M.D. and Svendsen, C. 2011. An assessment of the fate, behaviour and environmental risk associated with sunscreen TiO_2 nanoparticles in UK field scenarios. Science of the Total Environment 409(13): 2503–2510.

Johnston, A.M. 2013. Is the sacred for sale?: Tourism and indigenous peoples. Earthscan.

Kim, S. and Yoon, Y. (2003). The hierarchical effects of affective and cognitive components on tourism destination image. Journal of Travel and Tourism Marketing 14(2): 1–22.

Kirk, D. 2010. Environmental Management for Hotels. Taylor and Francis.

Lahens, L., Strady, E., Kieu-Le, T.C., Dris, R., Boukerma, K., Rinnert, E. and Tassin, B. 2018. Macroplastic and microplastic contamination assessment of a tropical river (Saigon River, Vietnam) transversed by a developing megacity. Environmental Pollution 236: 661–671.

Lange-Faria, W. and Elliot, S. (2012). Understanding the role of social media in destination marketing. Tourismos 7(1).

Li, Y.R., Lin, Y.C., Tsai, P.H. and Wang, Y.Y. (2015). Traveller-generated contents for destination image formation: Mainland China travellers to Taiwan as a case study. Journal of Travel & Tourism Marketing 32(5): 518–533.

Lira, Ivette. Los zoológicos de México deben dejar de ser espacios para la explotación para la explotación de animales: ONG's (sinembargo). (2017). https://www.sinembargo.mx/11-03-2017/3167235.

López, E., and Cendra, C. (2018). Big Data para optimizar las decisiones estratégicas de los destinos turísticos. Estudio de caso: Málaga. International Journal of Information Systems and Tourism (IJIST) 3(2): 67–78.

López, M.R. and López, M.R. (2017). PROGRAMA DE PUEBLOS MÁGICOS PERSPECTIVA, RETOS Y AVANCES. CASO CAPULÁLPAM DE MÉNDEZ. Revista Turydes: Turismo y Desarrollo, n. 23.

López, V.G., Moreno, L.R. and Meza, J.A. (2016). Assessment of sustainable tourism in Mexico. Global Journal of Business Research 10(3): 21–33.

Manning, E. (1996). Turismo: ¿dónde están los límites?

Navarro, D. (2015). Recursos turísticos y atractivos turísticos: conceptualización, clasificación y valoración. Cuadernos de turismo (35): 335–357.

Orgaz Agüera, F. and Moral Cuadra, S. (2016). El turismo como motor potencial para el desarrollo económico de zonas fronterizas en vías de desarrollo. Un estudio de caso. El periplo sustentable, (31).

PÉREZ, R.M.I. and VILLA, C.C. (2011). Teoría General del Turismo: un enfoque global y nacional. México: UABCS y Academia de Investigación Turística. Pag 54–75.

Pérez-Ramírez, C.A. and Antolín-Espinosa, D.I. (2016). Programa Pueblos Mágicos y desarrollo local: actores, dimensiones y perspectivas en El Oro, México. Estudios Sociales. Revista de alimentación contemporánea y desarrollo regional 25(47): 218–242.

Peterson, C. and Coats, J. 2001. Insect repellents-past, present and future. Pesticide Outlook 12(4): 154–158.

Pike, S. and Ryan, C. (2004). Destination positioning analysis through a comparison of cognitive, affective, and conative perceptions. Journal of Travel Research 42(4): 333–342.

Pike, S. and Page, S.J. (2014). Destination Marketing Organizations and destination marketing: A narrative analysis of the literature. Tourism Management 41: 202–227.

Paül i Agusti, D. (2019). La escasa representación turística de los ámbitos no urbanos. Una comparación de fuentes impresas e imágenes de Instagram.
Quesada, A.M., Rochel, R.B. and Valverde, F.L. (2018). Sistema inteligente para la planificación de visitas en destinos turísticos. In XII Congreso Internacional de Turismo y Tecnologías de la información y las comunicaciones (pp. 203-217). Universidad de Málaga (UMA).
Rábago, N.L.B. and Revah, L.O. (2000). El ecoturismo: ¿una nueva modalidad del turismo de masas? Economía, sociedad y territorio 2(7): 373–403.
Rivas, H. (1998). Los impactos ambientales en áreas turísticas rurales y propuestas para la sustentabilidad. Gestión turística (3): 47–75.
Rodríguez, R.G. (2010). Ecoturismo Mexicano: la promesa, la realidad y el futuro. Un análisis situacional mediante estudios de caso. El Periplo Sustentable (18): 37–67.
Ruiz Pelcastre, L.M. (2014). Aplicación del programa Pueblos Mágicos en Huasca de Ocampo. México: Universidad Naconal Autónoma de México.
San Martín, H. and Del Bosque, I.A.R. (2008). Exploring the cognitive–affective nature of destination image and the role of psychological factors in its formation. Tourism Management 29(2): 263–277.
Santos, I.R., Friedrich, A.C., Wallner-Kersanach, M. and Fillmann, G. 2005. Influence of socio-economic characteristics of beach users on litter generation. Ocean and Coastal Management 48(9-10): 742–752.
Saraniemi, S. and Kylänen, M. (2011). Problematizing the concept of tourism destination: An analysis of different theoretical approaches. Journal of Travel Research 50(2): 133–143.
SEMARNAT a. (2017). (Secretaria de Medio Ambiente y Recursos Naturales; Acciones y Programas, fomento ambiental) Turismo sustentable en México (Gobierno de México), 2017 https://www.gob.mx/semarnat/acciones-y-programas/turismo-sustentable-116615.
SEMARNAT b. (2017). Turismo Sustentable en México (Cuaderno de Divulgación Ambiental) http://biblioteca.semarnat.gob.mx/janium/Documentos/Ciga/Libros2013/CD002793.pdf.
Smith, P. (Ed.). (1998). The history of tourism: Thomas Cook and the origins of leisure travel (Vol. 4). Psychology Press.
Stovel, H. (1998). Risk preparedness: a management manual for world cultural heritage.
SUI-QUI, T. and LENG, H. (2015). Review on the development of a sustainability indicator system in agenda 21 for tourism in Mexico. International Review for Spatial Planning and Sustainable Development 3(2): 4–21.
Sumaila, U.R., Alder, J. and Keith, H. 2006. Global scope and economics of illegal fishing. Marine Policy 30(6): 696–703.
Taha, H. (1997). Urban climates and heat islands; albedo, evapotranspiration, and anthropogenic heat. Energy and Buildings 25(2).
Tassiopoulos, D. (Ed.). (2011). New Tourism Ventures: An Entrepreneurial and Mangerial Approach. Juta and Company Ltd.
Tiongson, J. (2015). Mobile app marketing insights: How consumers really find and use your apps. Retrieved from Think with Google: https://www.thinkwithgoogle.com/consumer-insights/mobile-app-marketing-insights/, 16th April 2020.
Torruco, M and Levy, S. (2019). (Gobierno de México). Resultados de la Actividad Turística Enero 2019. (Secretaria de Turismo; DATATUR; Subsecretaría de Planeación y Política Turística),2019. https://www.datatur.sectur.gob.mx/RAT/RAT-2019-01(ES).pdf.
Torruco, M. (Gobierno de México) Estrategia Nacional de Turismo 2019–2024 tendrá un sentido democrático (Secretaría de Turismo, Prensa). (2019). https://www.gob.mx/sectur/prensa/estrategia-nacional-de-turismo-2019-2024-tendra-un-sentido-democratico-miguel-torruco.
UNWTO. (2010). "UNWTO technical manual: Collection of Tourism Expenditure Statistics"_(PDF). World Tourism Organization. 1995. p. 10. Archived from the original (PDF) on 22 September 2010. Retrieved 26 April 2020.
Urry, J. (2012). Sociology beyond societies: Mobilities for the twenty-first century. Routledge.
Vázquez, J. Todos perdimos en caso Tajamar: FONATUR (El Economista). (2019). https://www.eleconomista.com.mx/estados/Todos-perdimos-en-caso-Tajamar-Fonatur-20190513-0001.html
Vigolo, V. (2017). Information and Communication Technologies: Impacts on Older Tourists' Behavior. In Older Tourist Behavior and Marketing Tools (pp. 85–104). Springer, Cham.
UNWTO, Sustainable Develope, Goals Sustainable Development, (2018) (https://www.unwto.org/sustainable-development).

World Tourism Organization. (2019). International Tourism Highlights, 2019 Edition, UNWTO, Madrid. (https://www.e-unwto.org/doi/book/10.18111/9789284421152 https://www.gob.mx/sectur/prensa/estrategia-nacional-de-turismo-2019-2024-tendra-un-sentido-democratico-miguel-torruco).

Warnken, J., Dunn, R.J. and Teasdale, P.R. 2004. Investigation of recreational boats as a source of copper at anchorage sites using time-integrated diffusive gradients in thin film and sediment measurements. Marine Pollution Bulletin 49(9-10): 833–843.

World Travel and Tourism Council, and APEC Tourism Working Group. (1998). The Economic Impact of Travel and Tourism Development in the APEC Region. World Travel and Tourism Council).

Yi, Y.K., del Vas, G.M. and Muñoz, A. (2018). Voyage 360: A holistic view on travel apps. In XII Congreso Internacional de Turismo y Tecnologías de la información y las comunicaciones (pp. 10–26). Universidad de Málaga (UMA).

Yorzinski, J.L., Chisholm, S., Byerley, S.D., Coy, J.R., Aziz, A., Wolf, J.A. and Gnerlich, A.C. (2015). Artificial light pollution increases nocturnal vigilance in peahens. PeerJ 3: e1174.

Zamorano Casal, F.M. (2002). Turismo alternativo: servicios turísticos diferenciados. Trillas turismo.

Zeng, S., Chiu, W., Lee, C.W., Kang, H.W. and Park, C. (2015). South Korea's destination image: comparing perceptions of film and nonfilm Chinese tourists. Social Behavior and Personality: an International Journal 43(9): 1453–1462.

Zhang, H., Zhang, W. and Wu, W. (2008, October). The mass customization of tourism products based on WEB2. 0: A collaboration model by both enterprises and tourists. In 2008 4th International Conference on Wireless Communications, Networking and Mobile Computing (pp. 1–5). IEEE.

Chapter 10

Whither Environmental Accounting. The Challenge of Ecological and Green Accounting

Danny García Callejas[1,*] and *Carmen A. Ocampo-Salazar*[2]

Introduction

The environment is a luxury that only rich countries can afford. Environmental Economics and Natural Resource Economics provide estimation of economic values for environmental services, non-market goods, renewable and non-renewable resources. Development requires polluting and depleting available resources in order to progress and create a wealthy society.

In this paper, we disagree with these statements and provide evidence of alternative methodologies and fields of accounting that incorporate social values as the center of analysis. While mainstream economics and accounting support their analyses on prices, heterodox theories include notions of ethics and the defense of life as obvious elements in the advance of society.

Green accounting introduces the tenets of social costs and social prices. The intent is to incorporate human needs and nature at the center of the decision-making process. Market failures provide a market-centric view, pretending that all elements of consideration are valued only in monetary terms. Decisions about life and the environment should account for emotions, sensations, perceptions, and value for future generations.

This paper aims to analyze critically the implications of Environmental Economics, Natural Resource Economics, Green Economics, and Ecological Economics on Environmental, Ecological, and Green Accounting Theory. By

[1] Economics Department, University of Antioquia, Calle 67 No. 53-108, room 13-106, Medellín, Colombia.

[2] Accounting Department, Eafit University, Carrera 49 No. 7 sur 50, room 26-110, Medellín, Colombia; cocampo2@eafit.edu.co

* Corresponding author: danny.garcia@udea.edu.co

deconstructing the dominant analytical notions of these fields, this study provides an understanding of the consequences of their use in society and organizations. Ecological and green accounting seem better suited for the defense of life and nature in the face of our current daunting environmental challenges. Finally, this research provides implications for research and practitioners.

Indeed, after discussing the main tenets of environmental, natural resource, ecological, and green economics, we present a brief literature review on green accounting and related fields. Then, the next section explains environmental, ecological, and green accounting in order to provide a comparison and understanding of the connections between economics and accounting. At the end, we conclude that green and ecological accounting are feasible and promising fields, providing alternatives to orthodox and functionalist propositions.

Economics, Natural Resources and the Environment

The term of environmental economics predates all other subfields studying economics and the environment. The Resources for the Future research institute coined the term in the 1950s (Pearce 2002). As Pearce (2002) explains, for this sub-discipline, environmental problems are a consequence of the economy failing to provide human welfare and the adequate combination of incentives to meet this collective goal.

However, Sandmo (2015) argues that environmental economics as a practice appeared more than two centuries ago. Accordingly, this author argues that the Marquis de Condorcet was the first to describe negative externalities while explaining the effects of pollution and the state's need to act over property rights to correct such a deviation. Similarly, Agnar Sandmo mentions classical economists such as David Ricardo, concerning their contributions to this young field.

Yet no theory is complete without an opposing perspective. Pearce (2002) mentions, then, ecological economics as a worthy adversary that criticizes some of the main assumptions of environmental economics. In particular, ecological economics bases its analysis on the fact that the earth's carrying capacity is limited and that inputs lack perfect substitutability as implied by the neoclassical approach. Additionally, ecological economists believe that policy actions should focus on quantities and not prices, that overconsumption leads to economic collapse, that efficiency does not imply optimality, and stress the need to focus on a human economy as both a social system and biophysical sphere (Gowdy and Erickson 2005).

In fact. ecological economics studies the interactions and co-evolution in time and space between ecosystems and human economies" (Xepapadeas 2009), understanding that the economy is a subsystem of the environment subject to the laws of physics. The environment is a complex system and, as such, rationality is seldom the explanation for the observed phenomena caused by humans.

In order to provide a more inclusive but orthodox alternative explanation to ecological economics, natural resource economics focuses on the allocation of natural resources, and the adequate human use of these resources, in order to provide sustainable actions and policies that ensure their availability, nature's protection, and an efficient economy.

Only in the early 20th century, economists explored the meaning and consequence of incorporating time and future generations in the allocation and availability of natural resources (Martinez-Alier 1995, Crocker 1999). Although natural resource economics was a preoccupation two centuries before, it was not until 1931 that Harold Hotelling would publish "The Economics of Exhaustible Resources", concluding that under perfect competition, natural resources would be exploited at a growth rate equivalent to the interest rate.

In contrast, the field of green economics considers this analysis incomplete. This field introduces principles from ecology, equity, social, political science, and environmental justice to better explain the implication in the use, consumption, and exploitation of nature's wonders (Kennet and Heinemann 2006). Additionally, it allows for the discussion on moral and ethical implications of the impact of human activity on nature.

Indeed, these four sub-disciplines complement and contradict each other. Their differences emerge by contrasting and comparing their assumptions. The focus on humans and the economic behavior or *homo economicus* rationale differ, as well as their object of analysis, feeding fields, conceptual approach, and general policy basis. Table 10.1 provides a summary and comparison of these aspects.

Table 10.1. Comparison of principles between environmental, natural resource, ecological, and green economics.

	Environmental Economics	**Natural Resource Economics**	**Ecological Economics**	**Green Economics**
Homo economicus	Rational agents and representative agent.	Rational agents and representative agent.	Economic agents immersed in a complex systems approach.	Agents with bounded rationality immersed in a social context.
Object of analysis	The effects of economic agents on the environment and the analysis of the impact of such policies.	Economic agents and their relation to environmental ecosystems. Natural endowments as natural constraints.	The ecosystem as a container for the economic system, in search of preserving the irreplaceable natural capital.	The ecosystem, its resilience and sustainability in relation to the promotion of life and economic activity.
Feeding fields	Economics and environmental science.	Economics and environmental science and biology.	Economics and environmental science, biology, and physics.	Economics and environmental science, biology, physics, social sciences, and political science.
Conceptual approach	Neoclassical economics. Institutionalism.	Neoclassical economics. Institutionalism.	Complex theory. Evolutionary economics. Econophysics.	Institutionalism. Structuralism.
Policy	Market policies. Market failure approach.	Market policies. Market failure approach.	Market policies. Market failure approach. State intervention and market regulation.	Market failure approach. State intervention and market regulation. Community action.

Source: Own elaboration, based on Managi and Kuriyama (2016).

Accordingly, each field has its own way of envisioning its research questions and problems. Environmental economics and natural resource economics use a positivist lens, with a quantitative approach, implying an objective analysis in their studies. In contrast, ecological economics allows also for an interpretivist focus, combining techniques drawn from physics, biology, and environmental sciences. This introduces, also, a subjective perspective.

In contrast, green economics uses an interpretivist and critical perspective, where qualitative techniques are also valid, as well as a subjective and contextual interpretation of the field's object of study. Nonetheless, the positivist approach and quantitative tools are welcome, on some occasions, to complement the explanation of the context at hand. Table 10.2 depicts these characteristics.

These four sub-disciplines of economics also have distinct differences on the determination of supply and demand quantities, scarcity, and availability of the resource. Different mechanisms, in each case, determine prices by some simple lack of a notion of equilibrium, a central concept in orthodox economics. Likewise, the understanding of economic growth and welfare are also diverse. Table 10.3 compares these differences.

Environmental economics and natural resource economics privilege market-oriented policies. Their conception of quantities, prices, and equilibrium implies their belief in competition and rational allocation of resources as coherent with Pareto efficiency and a utilitarian notion of welfare. These orthodox fields propose creating the conditions of a competitive economy to reach Pareto efficiency (Gowdy and Erickson 2005).

The goal of consumers in an Environmental or Natural Resource Economics framework is to maximize utility, subject to a budget constraint and of producers to maximize profit restricted to available technology and costs. Thus, environmental problems arise because of inadequate incentives, or market failures that are susceptible to correction via the same market (theorem of Coase) or shy and limited regulation and policy. The real problem is the lack of competitive conditions.

Table 10.2. Research approach, methods, and perspective.

Approaches	**Environmental Economics**	**Natural Resource Economics**	**Ecological Economics**	**Green Economics**
Epistemological approach	Positivist	Positivist	Positivist and interpretivist	Positivist, interpretivist, and critical
Research methods	Quantitative. Economic modelling and econometric analysis.	Quantitative. Economic modelling and econometric analysis.	Quantitative and qualitative. Physics, biology, and environmental science have a crucial role.	Quantitative and qualitative. Econometric analysis and contextual analysis.
Study perspective	Objective	Objective	Objective and subjective	Objective and subjective

Source: Own elaboration.

Table 10.3. Key concepts and conceptions in the four sub-disciplines that study the environment and economics.

Concept	Environmental Economics	Natural Resource Economics	Ecological Economics	Green Economics
Quantities and availability	Supply and demand determine quantities, subject to the characteristics of the environmental service and type of goods.	Reserves, availability, demand, and estimated current and future generations' demand and prices determine resource supply.	Reserves and intergenerational demand, supply, technological advances, uncertainty, and economic evolution determine quantities.	Intergenerational, social agreements, and preferences for nature and ecological balance determine availability and quantities.
Prices	Flexible and adjusted by market forces. Econometric methods estimate Non-market values.	Adjusted by market forces based on resource availability and reserves. Non-market values are estimated using econometric methods and mathematical modelling.	Prices must reflect complex systems conditions. Social prices should include resilience and environmental cost and natural resource depletion. Mathematical modelling allows estimating prices.	Social and environmental costs affect prices. Structural conditions affect prices of resources and environmental services. Combination of methods provides estimates.
Equilibrium	A general equilibrium and market imperfections framework may produce an equilibrium.	A market imperfections framework, endowments, available technology, and reserves provide a possible equilibrium.	Equilibrium is a path of constant adjustment influenced by a complex system. Equilibriums are rare, if they exist. Path dependency, altruism.	Disequilibria are permanent. Economic, social, political, and environmental variables alter prices, quantities, extraction, and output of natural resources and environmental goods and services.
Economic growth	Gross domestic product growth. Progress and a narrow view of economic development.	Sustainable development. Green growth.	Sustainable development. Humans as social actors.	Degrowth. Human scale development.
Notion of welfare	Utilitarian approach. Aggregation of preferences. Pareto efficiency.	Utilitarian approach. Aggregation of preferences. Pareto efficiency.	Equity, stability, resilience of environmental and social systems. Rawlsian approach.	Social welfare. Nature is a priority. Rawlsian approach. Environmental ethics.

Source: Own elaboration.

Social surplus—consumer and production surplus—gives way to cost benefit analysis as the main criteria for policy choice, in orthodox analyses. Nevertheless, Ecological and Green Economics use a multi-criteria decision analysis, as a representative tool for policy decision making (Doumpos and Zopounidis 2014, Gowdy and Erickson 2005, Xepapadeas 2009).

In growth terms, traditional branches of economics promote gross domestic product increase, although natural resource economics emphasizes the need for a sustainable growth. Indeed, an adequate management of resources and environmental products and services is called for in this vision. Yet, ecological economics and green economics claim the need for a human approach. The latter perspective is ready to sacrifice growth—or even stop it—to guarantee the protection of nature for current and future generations. The normal should be "to manage economics for nature as usual rather than to manage the environment for business as usual" (Kennet and Heinemann 2006).

Current debates led to the use of the term green economy, in order to express a path that the economy should follow in order to reach sustainability (Georgeson et al. 2017). As reviewed by Loiseau et al. (2016) and Georgeson et al. 2017, the green economy should be a term meaning transformation of the economy—with government intervention as a key element (Droste et al. 2016). Green economics contributes in this sense; however, the analyses of the concept and implications of the green economy are beyond the scope of this paper.

Nonetheless, green economics promotes an environmental ethics that fosters intergenerational justice. Such a view is compatible with a Rawlsian approach also, in part, shared by ecological economists. The idea of future generations enjoying the similar natural conditions as the current one requires a transition to an economy based on renewable resources (Tacconi 2017), and society's participation in the defense of the environment and nature (Collins and Barkdull 1995).

Accounting for the Environment

Three major groups can classify a review of the literature on environmental accounting: environmental and social accounting, green accounting, and ecological accounting. In the first group, some reviews show the complexity of these studies and locate their recent developments in Business Case, Evolutionary Theory, Critical or Radical theory, and Interpretivism or Critical Realism (Lehman and Chinthana Kuruppub 2017).

Criticism in this field emerges from the predominance of a managerial standpoint, abandoning environmental dilemmas in which organizations are involved. Altogether, five articles show discussions about this issue: Bebbington (1997), Mathews (1997), Gray (2002), Deegan and Soltys (2007), and Owen (2008). In addition, the reviews of Spence et al. (2010) and Parker (2011) delineate the main theoretical parameters of this field of study.

In accordance, Deegan (2002), Gray (2006, 2010), and Burritt and Schaltegger (2010), through legitimacy theory or contextual analysis, show that organizations use disclosure as a means for validating their interaction with the environment. As such, Milne and Gray (2013) show the disconnect between the practice of sustainability reporting and the protection of the environment, sometimes even resorting to facades (Cho et al. 2015) and camouflaging (Michelon et al. 2016).

In fact, there is a trend to use environmental accounting and reporting as a means of cultivating a positive company image with low disclosure quality, conclude Beck et al. (2010), by using a consolidated narrative interrogation method to organizations in Germany and Britain. Moreover, biodiversity reporting, like in the case of Sweden, is rare (Rimmel and Jonäll 2013). Organizations should improve their management decisionmaking to encourage sustainability (Burritt and Schaltegger 2010).

In relation to criticisms to this accounting domain and given the challenges that arise from the *Stockholm Declaration* of 1972, *Brundtland Report* of 1987, and *The Earth Summit* of 1992, other fields have begun to broaden discussions on environmental and social accounting, including green accounting and ecological accounting. In environmental accounting, there is a lack of plural and democratic views encouraging proposals of reform (Lehman 1999, Burritt and Schaltegger 2010); a business discourse dominates reporting (Spence 2007).

However, some authors indicate that studies on these issues have been scarce and the challenges are current (Gray and Laughlin 2012). Additionally, the few acknowledgments on environmental and social accounting, including green accounting and ecological accounting, are focused on financial accounting and its purposes (Owen et al. 2012, Thornton 2013). Thus, this gap in the literature justifies carrying out a study that differentiates among these accounting typologies.

The Environment in Accounting

Environmental Accounting

Environmental accounting interprets and incorporates valuation at simulated or current market prices of natural resources, commodities, and environmental services and goods in financial information and as an input for decision making. Such valuations are included in financial and accounting reports, representing the environment as capital.

The term, environmental accounting, is attributed to economist Peter Wood in the 1980s, although some of the first papers on the matter are those of Ullmann (1976) and Dierkes and Preston (1977). The former actually titles his paper "The corporate environmental accounting system", giving way to the development of this branch of accounting, a topic of interest for students, researchers, and practitioners (Fleischman and Schuele 2006).

Bebbington and Larrinaga (2014) point to the *Brundtland Report* of 1987 as the fuel that ignited the interest in the study of the environment from an accounting perspective. The report *The Greening of Accountancy: The Profession after Pearce* and special issues of *Accounting, Auditing and Accountability Journal* and *Accounting, Organizations and Society* in 1991 laid the basis for the development of this field (ibid.).

The tenets of environmental accounting reside in environmental economics and natural resource economics. The neoclassical framework lays the foundations for environmental accounting. Economic agents are rational in a perfect competition market structure without uncertainty nor market failures.

Rational agents imply an optimizing behavior, allowing for equilibria and an efficient use of available resources. Lack of uncertainty allows for discounting of future optimal decisions with complete markets. Thus, perfect information presumes knowledge of natural resource endowments and availability of environmental resources.

In accordance, market prices result from supply and demand forces. Such a setting allows prices to depict resource scarcity or depletion. An efficient solution means a consistent and systematic use and extraction of natural resources in accordance with current endowments. Hotelling's rule determines consumption and production patterns.

Market failures are plausible in this neoclassical framework. Externalities, tragedies of the common and public goods, are feasible and resolved using standard solutions. Coase's theorem, in a low transaction cost framework and by ruling on property rights, provides an adequate pattern for internalizing benefits or losses from positive or negative externalities.

Equivalent and artificial market prices serve as proxies of the value of natural resources and environmental goods and services without an explicit or existing market. The marginality principle allows equating, in principle, use value and exchange value. The configuration of quantities and prices is independent and excluded from political and social contexts.

Price formations are established equivalently worldwide, providing feasible and compatible values for global public goods. Limitations in price signalling are a consequence of international market barriers and excessive market interventions and regulations. Market liberalization and free mobility of capital, resources, and investment promote efficient use and allocation of resources.

This is compatible with a functionalist approach to accounting. Neoclassical economics serves as the rational framework for environmental economics and natural resource economics. These fields of economics provide the principles for environmental accounting. In fact, market prices and simulated market prices are the main input for accounting reports.

Asset, liability, and equity formation are possible through market valuations that provide the necessary information for financial reports. In fact, the environment, its by-products and services, interpreted as commodities, and natural resources are associated to natural capital (Mace 2019). This item is included in accounting reports and explained as any other component of the balance sheet.

Environmental accounting encourages analyses at the global, national, corporate, and managerial levels. Green accounting, in the orthodox perspective, is national environmental accounting (Bartelmus and Seifert 2018). This is one approach of green accounting. The following section depicts a heterodox view. Corporate environmental accounting concentrates its studies on environmental cost analyses and performance.

Some companies use corporate environmental reporting as a means for maintaining a positive image and increasing demand (Hopwood 2009) and, usually, reporting has a limited connection with sustainability (Gray 2010). In fact, Hopwood (2009), referring to carbon markets, puts in doubt that market transactions reported by corporations are coherent to environmentalists' demands.

Furthermore, corporate environmentalism provides an understanding on the way managers incorporate environmental and sustainable practices in their organizations; this influences the selection of measurements and tools (Banerjee 2002). In addition, environmental management accounting aims to facilitating decision-making. Nonetheless, corporate environmentalism is not the focus of this paper.

Consequences

Acknowledging the environment and its biological components and natural capital, resources, and commodities expresses an economic purpose. Capital is an essential part of production and essential in building profitability. Efficiency requires intensive and full usage of all available inputs, and natural capital is not an exception.

Market prices determine the rate of trade and exploitation of natural resources and environmental commodities. Depletion and extraction are a consequence of intensity use and price expectations. As goods and services, supply and demand determine availability, production, and consumption decisions. Private decisions dominate social demands.

Financial markets are incomplete, and time is an essential component that differentiates today's from tomorrow's commodities. Future and contingent markets emerge with natural resources, ignoring positive externalities and their public good nature. Private property and rational allocation exclude society from their enjoyment.

This framework ignores the role of nature as an interconnected system that allows the development of the economy and life. Furthermore, as an input, it is mainly involved in economic and financial decisions. The market ignores nature's characteristics as a biological system, provider of environmental balances, shelter for species and genetic stock.

In fact, in the era of the Anthropocene (Bebbington et al. 2019), telecoupling becomes a means to reconcile ecology, environment, accounting, and management. Bruckner et al. (2015) and Bebbington et al. (2019) provide a promising reflection inviting to tools beyond accounting and management to juxtapose the economic impacts on the environment and life. Indeed, prices in the economy fall short in accurately representing the negative consequences of production and consumption for society and nature.

Prices and Values are the Main Sources

Thus, environmental accounting relies on values and prices established through the market. Market methods, such as analysis of transactions data, residual valuation, and net income change allow for this computation. However, with the ecosystem and natural resources, such elements are usually unavailable and require being developed. Non-market methods include benefit cost analysis, travel cost, random utility, hedonic pricing, and contingent valuation. Table 10.4 describes these methods.

These methods assume that the economy works as described by orthodox and functionalist theories. Non-market methods seek to simulate market conditions, implying that individuals represent values, senses, desires, and feelings in monetary terms. These methods are incomplete for green and ecological accounting, in such a way that they ignore the importance of systemic and complex connections with the

Table 10.4. Market and non-market pricing and valuation methods.

Market methods	**Description**	**Non-market methods**	**Description**
Analysis of transaction data	The use of actual market prices on the transaction.	Benefit cost analysis	Compares total benefits to opportunity costs, ideally using market values for each considered component.
Residual valuation	Determines the price of intermediate natural goods by subtracting from the final price the associated costs for using the natural input.	Travel cost model	Estimates the implicit price of a natural asset or resource based on travel costs and invested time.
Net income change	Contribution of the natural resource to an organization's profits.	Random utility models	Valuations based on specific site attributes and extraordinary characteristics, from a subjective perspective.
		Hedonic pricing methods	Contrast of environmental sites differing on a single characteristic.
		Contingent valuation method	Estimates the value of good without a current existing market by using surveys.

Source: Own elaboration; based on Jensen and Bourgeron (2012) and Smith (1996).

ecosystem and the human and collective recognition that people give the environment and its wonders.

Ecological Accounting

Environmental accounting has evolved towards environmental financial accounting and environmental management accounting (Zhou et al. 2016). Ecological accounting, in contrast, is a branch of accounting that includes ecological principles and prioritizes the relationship between resources, economic activity, and the environment, in order to value and include such assets and equity in social and business planning.

Ecological concepts, measures, and values are at the core of ecological accounting (Birkin 2009). The need of integrating economic thinking and the environment and ecosystem, allows for an adequate accounting and decision-making process (Lei et al. 2014). Incorporating the stress on available resources and their resilience determines the true limits to sustainable growth and development.

Emergy and entropy are key concepts at the core of ecological accounting, in contrast to environmental accounting. Emergy is the direct and indirect use of energy required in the development of a product or service (Čuček et al. 2015, Le Corre 2016). This provides the connection with the biosphere. Thus, it implies the acknowledgement of the impact, origin, limits, and possibilities of economic activity from and through the environment (Alfsen and Greaker 2007, Lei et al. 2014).

Emergy or energy memory, or the energy required to produce another type of energy, is a term that was born in 1983 (Odum 2002). Its rationale allows for a framework that measures wealth in a holistic manner, since all activities require energy, transform energy, and use energy. Thus, it fosters a science-based method for evaluating economic activity, and for balancing the environment and the economy (Odum 1996).

Instead of calculating the ecological footprint, ecological accounting proposes determining the emergetic ecological footprint (Chen and Chen 2006, Peng et al. 2018). In addition, by including the environmental system as a whole, this enables understanding and measuring its connections with the economic system.

In accordance, diverse tools and methodologies complement the fundamental notion of emergy accounting. Ternary diagrams and plots are a recurrent tool in ecological accounting (Almeida et al. 2007). Patterson et al. (2017) argue for the need of finding common grounds and methods that co-learn from each other.

Ecological accounting challenges environmental accounting by incorporating the economy in the environmental ecosystem. Thus, prices, values, and measures of effects towards the biosphere or of use of natural equity and its services are included in economic activities.

The market is an incomplete institution antagonizing the environment, yet the environmental ecosystem is the sphere, including all human and non-human activities.

Tools for Environmental and Ecological Accounting

Although environmental and ecological accounting have a fundamental difference in the approach to value and measure the importance of the ecosystem in society, they share some accounting tools that allow them to apply their theory in practice. From the costing analysis perspective, material flow cost accounting and environmentally extended input-output analysis share similar principles and provide information consistent with both frameworks. Table 10.5 summarizes the meaning and deliverables of these and other practical tools.

The balanced scorecard is a performance measurement intended to balance financial, non-financial, short and long-term organizational goals and measures, derived from the firm's strategic planning (Hansen and Schaltegger 2016). Figge et al. (2003) describe this as a for-profit tool that considers the environmental and sustainability commitment of the organization.

In accordance, Hansen and Schaltegger (2016) and Journeault (2017) find that instrumental, political, and theoretical reasons encourage the inclusion of an environmental and sustainable perspective to the traditional balance scorecard. In fact, the growth of sustainability reporting is remarkable, fostered by legal, economic, and conventional reasons (Barker 2019).

Figge and Hahn (2005) develop a methodology to compute the cost of sustainability capital and sustainable value creation of companies. Kuosmanen and Kuosmanen (2009) provide an alternative and more accurate estimation—econometric strategy, although keeping the same principle; Kassem et al. (2016) develop an approach for small and medium sized business. In industry, Sustainable Value Stream Mapping is an important tool that evaluates manufacturing operations based on sustainability principles (Faulkner and Badurdeen 2014); nonetheless, this technique is beyond the scope of this paper.

Table 10.5. Accounting tools for the environment and the ecosystem.

Tool	Field	Reports	Meaning
Costing analysis			
Life Cycle Assessment	Ecological accounting	Sustainability reporting. The report includes physical and monetary metrics.	Determines the environmental burdens and impacts of a good or service at development, production, exchange, and consumption stages, and product or service decline or recycle.
Environmental Activity Based Costing	Environmental accounting	Cost and operational flow diagram, software model and report.	Technique applying active based costing to identify and measure direct environmental costs, separating indirect environmental costs. The environment is the key cost driver in this estimation and its application distinguishes between environment related costs normally attributed to joint environmental cost centers and environment driven costs, which can be direct, indirect, and contingent, and hidden in the general overhead.
Material Flow Cost Accounting	Ecological accounting and environmental accounting	Flow diagram with cost assignments to products, output, and waste. Material balance, with double entry, and assigned monetary value, including quantity centers.	Determines the possible environmental and financial effects of the organizations energy and material uses.
Environmentally extended input-output analysis	Ecological accounting and environmental accounting	Input-output matrix; production tree; trade, input and output flows and production layers; intensity vector, and consumption-based inventory vector.	Identifies production and consumption components and structures, expressed in physical quantities and monetary terms, establishing links between consumption and impacts.
Eco-balance	Ecological accounting	Report, flow diagrams, specified by product, process, or the organization as a whole.	Determines interior and exterior flows of resources, products, inputs, outputs, and waste in the organization.
Performance management			
Environmental Balance Scorecard	Environmental accounting	Balance scorecard and mapping.	Performance measurement aimed at balancing goals and measures with environmental corporate commitments derived from strategic planning.
Sustainability Balance Scorecard	Environmental accounting	Balance scorecard and mapping.	Performance measurement seeking tenable and balanced goals, based on strategic planning.

Table 10.5 Contd. ...

...*Table 10.5 Contd.*

Tool	Field	Reports	Meaning
Performance management			
Sustainable value	Environmental accounting and ecological accounting	Sustainable value tables, including opportunity costs and value contribution using performance data. Combines monetary and physical values—pollutants.	Measures and allows for the managing of sustainability performance, based on an opportunity cost framework, analyzing environmental, social, and economic contributions.

Source: Own elaboration, based on Hansen and Schaltegger (2016, 2018), Bartelmus and Seifert (2018).

Environmentally extended input-output analysis allows the identification of economic drivers of any environmental impact (Kitzes 2013).

Similarly, yet in a microeconomic perspective, Christ and Burritt (2017) provide an example of applying Material Flow Cost Accounting to the restaurant sector, related to the environmental management standard of the International Organization for Standardization, ISO, 14,051.

Green Accounting

National Accounts Perspective

This perspective aims to form a subaccount in the nation's gross domestic product, valuing the natural capital and its depletion in the economy. According to this approach, such an inclusion makes overt and clear the costs and benefits of the use of natural resources, the environment, and its products and services. Thus, society could make an efficient use of these resources, determining its financial and economic values.

However, such an exercise may be controversial, since depletion could already be included in the financial and accounting calculations made by extractive industries. Moreover, externalities, and their economic value, may result from an arbitrary determination and process. Simon (1981, 1998) argues that such a computational ordeal is futile since scarcity of resources fosters human invention and leads to cheaper prices of resources.

Nevertheless, economists and accountants develop this area of green accounting. Mäler (1991) and Hartwick (1990) propose including net changes in natural capital and deducting environmental damage. Several advocates agree on the need of a defined unit of account to do so, and Boyd and Banzhaf (2007) propose a methodology for doing so, as well as what should be included in such measurements (Boyd 2007). A need of connecting such computation to welfare is also clear (Dasgupta 2009).

Nonetheless, accountants were skeptical of these ideas until the 90s (Bebbington et al. 1994). According to Lamberton (2005), 2002 was the year of the accounting and the environment boom, although Ngwakwe (2012) argues that there is a "lack of standards, regulations and uniform accounting schemes". This topic, however, is beyond the scope and purpose of this paper.

Extended Approach

Green accounting has another side that includes the environmental, economic, and social issues combined. This is an attempt to provide an alternative and extended approach in comparison to environmental and ecological accounting. At the same time, this viewpoint tries to include social and environmental indicators, giving voice to the people, and providing importance to feelings, emotions, and sentiments related to the ecosystem.

Green economics provides the principles of this approach. Critical accountants have requested for an approach open to a democratic and plural process and avoiding a market centric vision (Gallhofer and Haslam 1997, Brown 2009). This field requires independence from mainstream and functionalist accounting, although the literature evidences a rapprochement between critical and orthodox scholars (Owen 2008). Brown and Dillard (2013) argue the need to move beyond managerial discourses.

In addition, as an accounting field, its worries are about the possibility of estimating values, although monetary ones are just a part of the analysis (Thornton 2013). In this regard, classifying and valuing goods and services is crucial. As such, private, public, club, common, merit and demerit goods are incorporated in the accounting analysis, since these have relations with environmental phenomena and issues.

In orthodox economics, an externality is the production and consumption of a good or service that affects the supply or demand of another good or service. Pollution is a recurrent example of a negative externality. In this case, these are demerit goods, whereas positive externalities are included in merit goods. Public goods, or non-rival and non-excluding, are local or global. This is the case for environmental goods and services, such as clean air. Common goods are rival, but non-excluding goods, such as lakes. In addition, club goods are excluding but non-rival, in contrast with private goods that are excluding and rivalrous.

Table 10.6 provides the alternative prices that green accounting identifies as inclusive and complete in comparison to the orthodox solution or dominant valuation.

Social costs and social prices, applied to the environment, are derived from Kapp's theory (Berger 2008), a heterodox approach in economics to pricing. Similarly, environmental pricing and social prices are a result of more than market forces (Hannon 2001). In this case, society's valuation, natural elements benefits, and environmental values are included. Table 10.7 describes the environmental values included in the traditional approach and the alternative prescription proposed by green accounting.

Similarly, costs are an important issue for organizations and society. Fitcher et al. (1997, cited by Schaltegger and Burritt 2017) classifies environmental costs in four groups: separate computations, full cost, direct costing, and process costing. Nonetheless, Table 10.8 provides a description of ecosystem costs, including a green account approach with an environmental and non-financial perspective.

Environmental restoration, depletion, and resilience require external information to the organization. For the other cost types, the organization relies on available and accurate information that may improve by combining it with additional data from different sources. The key element in green accounting is the plural and democratic

Table 10.6. Perspective of green accounting on prices and solutions for environmental problems.

Type of good	Dominant valuation	Market structure	Economic accounting approach	Green accounting implication
Private goods	Market prices	Competition and monopolistic competition	Economic analytical approach	Social costs and social prices
Public goods (global public goods and local public goods (Tibeout))	Lindhal prices	Marked up prices	Economic Valuation	Environmental pricing and social costs
Club goods	Market prices	Imperfect competition	Economic analytical approach	Environmental pricing
Common goods	Pigouvian (tax) prices	Imperfect competition	Economic Valuation	Environmental pricing
Merit goods	Market prices with mark-up	Monopolistic competition	Positive externalities approach	Environmental pricing and social benefits
Demerit goods	Market prices with mark-up	Monopolistic competition	Negative externalities approach	Environmental pricing and social costs

Source: Own elaboration; based on Streeten (1997), Managi and Kuriyama (2016).

Table 10.7. Environmental values and green accounting alternatives.

Revenue, asset or equity type	Description	Green accounting alternative
Resource availability and natural capital appreciation	Current and future revenue derived from the exploitation, value and availability of natural resources.	Intergenerational value. Importance as social, historical, and cultural equity.
Entertainment and natural services	Nature provides a means for human contemplation of the wonders of life. Sightseeing and sports practicing are just some of the examples.	The ecosystem includes society and the economy.
Environmental balancing	Nature allows keeping in balance and minimizing the effect of disasters and human externalities.	Acknowledgment of a complex system. Assessment of the benefits of nature. Nature as the priority.
Non-use values: existence value	Valuation given to nature, environmental sites just due to their existence.	Nature as the source of life. Biological and chemical values.
Non-use value: bequest value	Value given by the current generation of the availability of the resource for future generations.	Intergenerational value; importance of species and the ecosystem.

Source: Own elaboration; based on Hecht (2012).

Table 10.8. Cost type and green accounting approach.

Cost type	Description	Green accounting approach
Pollution Prevention	Costs incurred to prevent air, soil, underground, and water pollution along with water treatment facilities, soil conservation activities, and natural reserves and parks protection.	Social justice and inclusion of community and affected stakeholders. Analysis of carbon dioxide emissions in tons. Carbon dioxide equivalents. Chlorofluorocarbons equivalent emissions in tons.
Environmental Protection	Costs of energy saving measures, global warming reduction actions, and climate change measures and species protection.	Interviews and energy saving, climate change, and species protection policies. Analysis of carbon dioxide emissions in tons. Direct and indirect energy consumption, energy from renewable sources, percentage of consumption reduction, and number, assessment, and value of awareness campaigns to protect species and about climate change.
Resource Recycling	Costs incurred in waste reduction and disposal as well as for water conservation, rainwater usage, recollection, and management of products and other strategies seeking an efficient resource usage.	Life history of the people involved in the primary process of recycling. Pro-environmental behavior campaigns. Interviews to employees in charge of the recycling process. Indicators on: reuse of resources in the organization, savings in physical and economic values, percentage of recycled inputs.
Environmental Restoration	Cost of environmental restoration operations (eliminating soil and ground water contamination, environmental compensation, community social fabric restoration).	Interviews and focus groups. Identification of historical and cultural recognition and losses. Participatory action research. Indicators on: percentage of recovery and loss, physical and monetary; percentage and number of communities protected or compensated.
Management	Management-related environmental protection costs including environmental promotion activities and costs associated with acquiring and maintaining certifications.	Policies and programs associated with protection of the environment. Organizational efforts including staff and employees, properly documented. Organizational ethnography, identifying pro-environmental practices, behaviors, and their impact on employees and staff. Value of promotion activities and certifications.
Social Promotion Activities	Environmental protection costs stemming from participation in social activities such as participation in organizations concerning with environmental preservation, among others.	Documentary review, discourse analysis of company statements. Value of support to green organizations.

Table 10.8 Contd. ...

...Table 10.8 Contd.

Cost type	Description	Green accounting approach
Research and Development	Environmental protection costs for research and development activities and costs of environmental solutions business activities (Green product and environmental technology design and development costs, environmental solutions business costs, others), etc.	Organization's statement on the use of green inputs.
Depletion and Resilience	Economic value of the loss of natural capital and cost of natural resource or environmental recovery.	Environmental compensation to affected sites and communities. Environmental impact analysis of the organization's projects.

Source: Own elaboration; based on Hecht (2012) and Jasch (2008).

nature of the costing process. The voice of the community, and society, with life at the center of the analysis is crucial.

Research Implications

Green and ecological accounting promote the need to acknowledge direct and indirect energy costs. Emergy, this modern concept, should reshape private and public accounting by internalizing feasible energy availability, actual costs, and externalities. This means that accounting reports must integrate a social price perspective. Market price value may undervalue the costs for society through environmental degradation.

Integrated reporting shows little impact in transforming polluting actions by firms that use sustainability reports as a means for creating a hyperreal context (Rodríguez-Gutiérrez et al. (2019), Boiral and Heras-Saizarbitoria (2019)). As explained by Bebbington et al. (2019), organizations must evolve with the context, allowing for accounting to foster accountability in light of the Anthropocene and using conceptual tools that transcend management and accounting research.

Indeed, the use of big data analytics in accounting for environmental and sustainability purposes provides a promising ground for research. Auditing, forensic accounting, tax accounting and organizational control are fields of management and accounting that will benefit from this approach. The link with consumers and stakeholders, plus the gains from transparency, allow multiple sources of information to enrich standard and future accounting reporting.

Ecological and green accounting will benefit from the multiple use of these tools and the inclusion of Anthropocene discussions. This may induce social and environmental reporting to reshape towards a responsible and independent practice that defends pluralism and communities, and not a means for silencing and obscuring extractivist and bigoted practices (Journeault et al. 2020). Telecoupling is a probable pathway for practitioners (Bruckner et al. 2015, Bebbington et al. 2019), and a feasible strategy to integrate social, economic, environmental, ecological, and cultural perspectives in accounting and management.

Conclusions

Environmental and natural resources economics are the basis for environmental accounting. The latter requires information on prices or estimations strategies, allowing such to be available. These branches of economics use the neoclassical framework to determine prices and quantities and explain values in market and non-market settings. Consistent with this framework, market solutions act as artificial corrections of market failures. The goal of this analysis is to develop equilibrium prices and foster efficiency.

This positivist approach is consistent with a financial analysis of the environment and biosphere. Thus, accounting translates the economic information into resources that are objects of efficient allocation by society and organizations. In this regard, economic and financial purposes prevail ignoring the context of the economy as a subsystem of the ecosystem and the emotions, feelings, and historic and cultural equity associated with the environment.

Ecological economics contributes the fundamentals of ecological accounting. This approach is an alternative to orthodox structure, by understanding the economy as dependent on the environment. As such, this view considers the ecosystem as a complex system that includes society and the economy. Emergy and entropy are key concepts in establishing this connection and enabling the use of practical tools such as: Life Cycle Assessment, Material Flow Cost Accounting, Environmentally Extended Input-Output Analysis, Eco-Balance, and Sustainable Value.

As ecological accounting, green accounting is a challenge to environmental accounting. This approach includes in its analyses the human perspective and notion of democracy, pluralism, and historical and cultural equity. Simultaneously, the embracement of these themes encourages a comprehensive approach in coherence with ecological accounting, although awarding more emphasis to the value of life in its explanations and tools.

Social prices and environmental pricing are alternatives to market prices proposed by environmental accounting. Green accounting provides alternatives that consider life, species, and biodiversity as central in valuing the ecosystem in contrast to the orthodox view that considers environmental values and cost types. Financial and economic goals are secondary for green accounting, although included in this framework. Consequently, green accounting, a critical perspective, establishes itself as a challenge to traditional financial reckoning of the environment.

Acknowledgments

We are thankful for the financial support from the Universidad de Antioquia, Fondo Patrimonial de Vicerrectoría de Docencia, Facultad de Ciencias Económicas, Universidad Eafit and Universidad del Valle. Research project 2019-25892.

References

Alfsen, K.H. and Greaker, M. (2007). From natural resources and environmental accounting to construction of indicators for sustainable development. Ecological Economics 61(4): 600–610.

Almeida, C.M.V.B., Barrella, F.A. and Giannetti, B.F. (2007). Emergetic ternary diagrams: five examples for application in environmental accounting for decision-making. Journal of Cleaner Production 15(1): 63–74.

Banerjee, S.B. (2002). Corporate environmentalism: the construct and its measurement. Journal of Business Research 55(3): 177–191.

Barker, R. (2019). Corporate natural capital accounting. Oxford Review of Economic Policy 35(1): 68–87.

Bartelmus, P. and Seifert, E.K. (2018). Green Accounting, Routledge, New Jersey.

Bebbington, J., Gray, R., Thomson, I. and Walters, D. (1994). Accountants' attitudes and environmentally-sensitive accounting. Accounting and Business Research 24(94): 109–120.

Bebbington, J. (1997). Engagement, education and sustainability: A review essay on environmental accounting. Accounting, Auditing & Accountability Journal 10(3): 365–381.

Bebbington, J. and Larrinaga, C. (2014). Accounting and sustainable development: An exploration. Accounting, Organizations and Society 39(6): 395–413.

Bebbington, J., Österblom, H., Crona, B., Jouffray, J.-B., Larrinaga, C., Russell, S. and Scholtens, B. (2019). Accounting and accountability in the Anthropocene. Accounting, Auditing & Accountability Journal 33(1): 152–177.

Beck, C., Campbell, D. and Shrives, P.J. (2010). Content analysis in environmental reporting research: Enrichment and rehearsal of the method in a British German context. The British Accounting Review 42(3): 207–222.

Berger, S. (2008). K. William Kapp's theory of social costs and environmental policy: Towards political ecological economics. Ecological Economics 67(2): 244–252.

Birkin, F. (2009). Ecological accounting: new tools for a sustainable culture. International Journal of Sustainable Development & World Ecology 10(1): 49–61.

Boiral, Olivier and Heras-Saizarbitoria, Iñaki. (2019). Sustainability Reporting Assurance: Creating Stakeholder Accountability Through Hyperreality? Journal of Cleaner Production 243.

Boyd, J. (2007). Nonmarket benefits of nature: What should be counted in green GDP? Ecological Economics 61(4): 716–723.

Boyd, J. and Banzhaf, S. (2007). What are ecosystem services? The need for standardized environmental accounting units. Ecological Economics 63(2-3): 616–626.

Brown, J. (2009). Democracy, sustainability and dialogic accounting technologies: Taking pluralism seriously. Critical Perspectives on Accounting 20(3): 313–342.

Brown, J. and Dillard, J. (2013). Agonizing over engagement: SEA and the 'death of environmentalism' debates. Critical Perspectives on Accounting 24(1): 1–18.

Bruckner, Martin, Fischer, Günther, Tramberend, Sylvia and Giljum, Stefan. (2015). Measuring telecouplings in the global land system: A review and comparative evaluation of land footprint accounting methods. Ecological Economics 114(June 2015): 11–21.

Burritt, R.L and Schaltegger, S. (2010). Sustainability accounting and reporting: fad or trend? Accounting, Auditing & Accountability Journal 23(7): 829–846.

Chen, B. and Chen, G.Q. (2006). Ecological footprint accounting based on emergy—A case study of the Chinese society. Ecological Modelling 198(1-2): 101–114.

Cho, C.H., Laine, M., Roberts, R.W. and Rodrigue, M. (2015). Organized hypocrisy, organizational façades, and sustainability reporting. Accounting, Organizations and Society 40(1): 78–94.

Christ, K. and Burritt, R. (2017). Material flow cost accounting for food waste in the restaurant industry. British Food Journal 119(3): 600–612.

Collins, D. and Barkdull, J. (1995). Capitalism, Environmentalism, and Mediating Structures. Environmental Ethics 17(3): 227–244.

Crocker, T.D. (1999). A short history of environmental and resource economics. pp. 32–47. *In*: van den Bergh, J.C.J.M. (ed.). Handbook of Environmental and Resource Economics. Edward Elgar Publishing, New York.

Čuček, L., Klemeš, Jiri and Kravanja, Zdravko. (2015). Chapter 5 - Overview of environmental footprints. pp. 131–193. *In*: Assessing and Measuring Environmental Impact and Sustainability, Publisher: Butterworth-Heinemann, Editors: Jiří Jaromír Klemeš.

Dasgupta, P. (2009). The Welfare Economic Theory of Green National Accounts. Environmental and Resource Economics 42(1): 3–38.

Deegan, C. (2002). Introduction: The legitimising effect of social and environmental disclosures—a theoretical foundation. Accounting, Auditing & Accountability Journal 15(3): 282–311.

Deegan, C. and Soltys, S. (2007). Social accounting research: An Australasian perspective. Accounting Forum 31(1): 73–89.

Dierkes, M. and Preston, L.E. (1977). Corporate social accounting and reporting for the physical environment: A critical review and implementation proposal. Accounting, Organizations and Society 2(1): 3–22.

Doumpos, M. and Zopounidis, C. (2014). An overview of multiple criteria decision aid. pp. 11–22. *In*: Doumpos, M. and Zopounidis, C. (eds.). Multicriteria Analysis in Finance. Springer Briefs in Operations Research, Springer, Amsterdam.

Droste, N., Hansjürgens, B., Kuikman, P., Otter, N., Antikainen, R., Leskinen, P., Pitkänen, K., Saikku, L., Loiseau, E. and Thomsen, M. (2016). Steering innovations towards a green economy: Understanding government intervention. Journal of Cleaner Production 135: 426–434.

Faulkner, W. and Badurdeen, F. (2014). Sustainable Value Stream Mapping (Sus-VSM): methodology to visualize and assess manufacturing sustainability performance. Journal of Cleaner Production 85: 8–18.

Figge, F., Hahn, T., Schaltegger, S. and Wagner, M. (2003). The sustainability balanced scorecard as a framework to link environmental management accounting with strategic management. *In*: Bennett, M., Rikhardsson, P.M. and Schaltegger, S. (eds.). Environmental Management Accounting—Purpose and Progress. Eco-Efficiency in Industry and Science, vol 12. Springer, Dordrecht.

Figge, F. and Hahn, T. (2005). The cost of sustainability capital and the creation of sustainable value by companies. Journal of Industrial Ecology 9(4): 47–58.

Fleischman, R.K. and Schuele, K. (2006). Green accounting: A primer. Journal of Accounting Education 24(1): 35–66.

Gallhofer, S. and Haslam, J. (1997). The direction of green accounting policy: critical reflections. Accounting, Auditing & Accountability Journal 10(2): 148–174.

Georgeson, L., Maslin, M. and Poessinouw, M. (2017). The global green economy: a review of concepts, definitions, measurement methodologies and their interactions. Geo: Geography and Environment 4(1): 1–23.

Gowdy, J. and Erickson, J.D. (2005). The approach of ecological economics. Cambridge Journal of Economics 29(2): 207–222.

Gray, R. (2002). The social accounting project and Accounting Organizations and Society. Privileging engagement, imaginings, new accountings and pragmatism over critique? Accounting, Organizations and Society 27(7): 687–708.

Gray, R. (2006). Social, environmental and sustainability reporting and organisational value creation?: Whose value? Whose creation? Accounting, Auditing & Accountability Journal 19(6): 793–819.

Gray, R. (2010). Is accounting for sustainability actually accounting for sustainability… and how would we know? An exploration of narratives of organisations and the planet Accounting, Organizations and Society 35(1): 47–62.

Gray, R., and Laughlin, R. (2012). It was 20 years ago today: Sgt Pepper, Accounting, Auditing & Accountability Journal, green accounting and the Blue Meanies. Accounting, Auditing & Accountability Journal 25(2): 228–255.

Hannon, B. (2001). Ecological pricing and economic efficiency. Ecological Economics 36(1): 19–30.

Hansen, E.G. and Schaltegger, S. (2016). The Sustainability Balanced Scorecard: A Systematic Review of Architectures. Journal of Business Ethics 133(2): 193–221.

Hansen, E.G. and Schaltegger, S. (2018). Sustainability Balanced Scorecards and their Architectures: Irrelevant or Misunderstood? Journal of Business Ethics 150(4): 937–952.

Hartwick, J.M. (1990). Natural resources, national accounting and economic depreciation. Journal of Public Economics 43(3): 291–304.

Hecht, J.E. (2012), National Environmental Accounting: Bridging the Gap between Ecology and Economy, Routledge, New Jersey.

Hopwood, A.G. (2009). Accounting and the environment. Accounting, Organizations and Society 34(3-4): 433–439.

Jasch, C.M. (2008). Environmental and Material Flow Cost Accounting: Principles and Procedures. Volume 25 of Eco-Efficiency in Industry and Science, Springer Science & Business Media, New York.

Jensen, M.E. and Bourgeron, P.S. (2012), A Guidebook for Integrated Ecological Assessments. Springer Science & Business Media, New York.

Journeault, M. (2017). The sustainability balanced scorecard: a systematic review of architectures. Social and Environmental Accountability Journal 37(1): 78–79.

Journeault, Marc, Levant, Yves and Picard, Claire-France. (2020). Sustainability performance reporting: A technocratic shadowing and silencing. Critical Perspectives on Accounting, In Press.

Kassem, E., Trenz, O., Hřebíček, J. and Faldik, O. (2016). Sustainability assessment using sustainable value added. Procedia – Social and Behavioral Sciences 220: 177–183.

Kennet, M. and Heinemann, V. (2006). Green Economics: setting the scene. Aims, context, and philosophical underpinning of the distinctive new solutions offered by Green Economics. International Journal of Green Economics 1(2): 68–102.

Kitzes, J. (2013). An introduction to environmentally-extended input-output analysis. Resources 2(4): 489–503.

Kuosmanen, T. and Kuosmanen, N. (2009). How not to measure sustainable value (and how one might). Ecological Economics 69(2): 235–243.

Lamberton, G. (2005). Sustainability accounting—a brief history and conceptual framework. Accounting Forum 29(1): 7–26.

Le Corre, Olivier. (2016). Emergy, Elsevier, Amsterdam.

Lehman, G. (1999). Disclosing new worlds: a role for social and environmental accounting and auditing. Accounting, Organizations and Society 24(3): 217–241.

Lehman, G. and Chinthana Kuruppub, S. (2017). A framework for social and environmental accounting research. Accounting Forum 4(3): 136–146.

Lei, K., Zhou, S. and Wang, Z. (2014). Ecological Energy Accounting for a Limited System: General Principles and a Case Study of Macao, Springer, Berlin.

Loiseau, E., Saikku, L., Antikainen, R., Droste, N., Hansjürgens, B., Pitkänen, K., Leskinen, P., Kuikman, P. and Thomsen, M. (2016). Green economy and related concepts: An overview. Journal of Cleaner Production 139: 361–371.

Mace, G.M. (2019). The ecology of natural capital accounting. Oxford Review of Economic Policy 35(1): 54–67.

Mäler, K.G. (1991). National accounts and environmental resources. Environmental and Resource Economics 1(1): 1–15.

Managi, S. and Kuriyama, K. (2016), Environmental Economics. Volume 17 de Routledge Textbooks in Environmental and Agricultural Economics, Taylor & Francis, New York.

Martinez-Alier, J. (1995). Distributional issues in ecological economics. Review of Social Economy 53(4): 511–528.

Mathews, M.R. (1997). Accounting research: Is there a silver jubilee to celebrate? Accounting, Auditing & Accountability Journal 10(4): 481–531.

Michelon, G., Pilonato, S., Ricceri, F. and Roberts, R.W. (2016). Behind camouflaging: traditional and innovative theoretical perspectives. Sustainability Accounting, Management and Policy Journal 7(1): 2–25.

Milne, M.J. and Gray, R. (2013). W(h)ither Ecology? The triple bottom line, the global reporting initiative, and corporate sustainability reporting. Journal of Business Ethics 118(1): 13–29.

Ngwakwe, C.C. (2012). Rethinking the accounting stance on sustainable development. Sustainable Development 20(1): 28–41.

Odum, H.T. (1996), Environmental Accounting: Emergy and Environmental Decision Making, John Wiley & Sons, New York.

Odum, H.T. (2002). Emergy accounting. pp. 135–146. *In*: Bartelmus, P. (ed.). Unveiling Wealth, Springer, Dordrecht.

Owen, D. (2008). Chronicles of wasted time?: A personal reflection on the current state of, and future prospects for, social and environmental accounting research. Accounting, Auditing & Accountability Journal 21(2): 240–267.

Owen, D., Gray, R. and Bebbington, J. (2012). Green accounting: Cosmetic irrelevance or radical agenda for change? Asia-Pacific Journal of Accounting 4(2): 175–198.

Parker, L.D. (2011). Twenty-one years of social and environmental accountability research: A coming of age. Accounting Forum 35(1): 1–10.

Patterson, M., McDonald, G. and Hardy, D. (2017). Is there more in common than we think? Convergence of ecological footprinting, emergy analysis, life cycle assessment and other methods of environmental accounting. Ecological Modelling 362(C): 19–36.

Pearce, D. (2002). An intellectual history of environmental economics. Annual Review of Energy and the Environment 27: 57–81.

Peng, W., Wang, X., Li, X. and Chenchen, H. (2018). Sustainability evaluation based on the emergy ecological footprint method: A case study of Qingdao, China, from 2004 to 2014. Ecological Indicators 85(2): 1249–1261.

Rimmel, G. and Jonäll, K. (2013). Biodiversity reporting in Sweden: corporate disclosure and preparers' views. Accounting, Auditing & Accountability Journal 26(5): 746–778.

Rodríguez-Gutiérrez, P., Correa, C. and Larrinaga, C. (2019). Is integrated reporting transformative? An exploratory study of non-financial reporting archetypes. Sustainability Accounting, Management and Policy Journal 10(3): 617–644.

Sandmo, A. (2015). The early history of environmental economics. Review of Environmental Economics and Policy 9(1): 43–63.

Schaltegger, S. and Burritt, R.L (2017). Contemporary Environmental Accounting: Issues, Concepts and Practice, Routledge, New Jersey.

Simon, J.L. (1998). The Ultimate Resource 2, Princeton University Press, Ney Jersey.

Simon, Julian Lincoln. (1981). The Ultimate Resource, Princeton University Press, Ney Jersey.

Smith, V.K. (1996). Estimating Economic Values for Nature: Methods for Nonmarket Valuation, Edward Elgar Publishing, New York.

Spence, C. (2007). Social and environmental reporting and hegemonic discourse. Accounting, Auditing & Accountability Journal 20(6): 855–882.

Spence, C., Husillos, J. and Correa-Ruiz, C. (2010). Cargo cult science and the death of politics: A critical review of social and environmental accounting research. Critical Perspectives on Accounting 21(1): 76–89.

Streeten, P.P. (1997). Thinking about Development. Raffaele Mattioli Lectures on the History of Economic Thought, Cambridge University Press, Cambridge.

Tacconi, L. (2017). Biodiversity and Ecological Economics: Participatory Approaches to Resource Management, Routledge, New York.

Thornton, D.B. (2013). Green accounting and green eyeshades twenty years later. Critical Perspectives on Accounting 24(6): 438–442.

Ullmann, A.E. (1976). The corporate environmental accounting system: a management tool for fighting environmental degradation. Accounting, Organizations and Society 1(1): 71–79.

Xepapadeas, A. (2009). Ecological economics: principles of economic policy design for ecosystem management. pp. 740–747. *In*: Levin, S. (ed.). The Princeton Guide to Ecology, Princeton University Press, New Jersey.

Zhou, Z.F., Ou, J. and Li, S.H. (2016). Ecological accounting: a research review and conceptual framework. Journal of Environmental Protection 7(5): 643–655.

CHAPTER 11

One Health for the New Era of our Common Oikos

Maria Gabriela Valle Gottlieb,[1,*] *Vilma Maria Junges*,[2]
Raquel Seibel[3] and *Vera Elizabeth Closs*[4]

Introduction

A Brief History of Us All: From the Big Bang to *Homo sapiens*

The most accepted theory for the mysterious event that created the Universe is the Big Bang. This theory postulates the creation of each atom, star, galaxy, and all our history occurred from a minimal portion of highly condensed energy. Thus, in a very high temperature, it entered the process of inflation or expansion. Several researchers around the world have suggested in the beginning the four fundamental forces of the universe (gravitational, electromagnetic, strong nuclear, and weak nuclear) were densely compressed at a high temperature in a single point. Therefore, for some reason (still unknown), it began to expand and transform. As this point expanded (considered the smallest fraction of the atom), the universe originated as the density of matter decreased (as it spreads), and the radiation density decreased even more rapidly. Besides spreading, this occurred since the universe temperature was decreasing simultaneously. In a hundredth of a second, an effervescent set of quarks, electrons, and other particles collided, cooling rapidly to form protons and

[1] Biomedical Gerontology Program of School of Medicine and Institute of Geriatrics and Gerontology Pontifical Catholic University of Rio Grande do Sul, Porto Alegre, Brazil; Av. Ipiranga, 6681/703, Porto Alegre/RS, Brazil.

[2] Integrated Obesity Treatment Center, Porto Alegre/RS, Brazil;Rua José Scutari, 371; Porto Alegre/RS, Brazil; vjungesnutri@yahoo.com.br

[3] Biomedical Gerontology Program of School of Medicine and Institute of Geriatrics and Gerontology Pontifical Catholic University of Rio Grande do Sul, Porto Alegre, Brazil; Rua Demétrio Ribeiro, 943/308. Porto Alegre/RS, Brazil; raquelseibel27@gmail.com

[4] Study Group on Cardiometabolic Risk, Aging and Nutrition, Institute of Geriatrics and Gerontology, Pontifical Catholic University of Rio, Grande do Sul (IGG-PUCRS), Porto Alegre/RS, Brazil; Tv. Coronel Antonio Carneiro Pinto, 165/501, Porto Alegre/RS, Brazil; veraec@terra.com.br

* Corresponding author: vallcgot@hotmail.com

neutrons. In the following minutes, electrons combined with protons and neutrons at very high speed created atoms and molecules, such as hydrogen and helium, thus the light could finally shine (Gamow 1948). From that first, mysterious, fundamental, and complex moment, any mineral, vegetable, or animal form became possible. According to astrophysical and geological research, this first extraordinary event happened approximately 13.82 billion years ago. The appearance of the Solar System and Planet Earth occurred about 4.5 billion years ago, and the first hominid appeared around 6 million years ago (Futuyama 2009) (Table 11.1).

From this evidence, the Universe (vast ecosystem) was found as an open and dynamic system, presenting a non-linear behavior, sensitive to both internal and external conditions (Oestreicher 2007). Thus, it is also sensitive to environmental

Table 11.1. Some milestones in the history of the Earth and living beings.

Geological era Period	Time (years)	Historic mark	Evolution of living beings
Precambrian	~ 4600	**Origin of the Earth**	Formation of the Earth's crust. 4.2 billion years ago: formation of the first seas. Atmosphere without oxygen.
Archeozoic	~ 3960	**Origin of living beings**	First prokaryotic cells (arquibacteria and algae in the seas).
Proterozoic	2500		First single-celled eukaryotic beings.
	950–2,0		First multicellular organisms.
Paleozoic			
Cambrian	543		First arthropods (trilobites).
Ordovician	500		First vertebrate organisms.
Devonian	408		First amphibians and insects.
Carboniferous	354		Amphibians originating reptiles. Extensive trees and forests.
Permian	290		Diversification of reptiles, extinction of many marine invertebrates.
Mesozoic			
Triassic	250		Appearance of dinosaurs.
Cretaceous	144		First placental mammals.
Cenozoic			
Paleocene	65		First primates, modern birds. Mass extinctions of dinosaurs.
Pliocene	5,2		First ancestors of humans.
Holocene	0,01		Neolithic. Age of copper, bronze, and iron. Early civilizations and the emergence of writing.
Present	2020	**Modern man**	Accurate scientific and technological knowledge. Contamination, destruction, deforestation of biomes and ecosystems accelerating global warming. Life on Earth in danger.

Source: Dalgalarrondo 2011, Futuyama 2009.

conditions and interactions or stimuli that can disturb or generate disproportionate responses to a given initial condition (Holm 2002). Above all, nature itself and biological systems present on the one hand, characteristics of order and stability, emphasizing the concept of homeostatic balance, and on the other, chaotic behavior (disorder, irregularity, unpredictability, non-linearity) (Sharma 2009). However, chaotic behavior is not synonymous with random and erratic movement, it is deterministic and standardized (Capra 1996). Since chaotic nonlinear systems are sensitive to a given initial condition, any change in the state system (the smallest fraction of the atom, a cell, an organism, an ecosystem, or even a meteor path) can trigger large-scale consequences over time. This is known as the "Butterfly Effect", described by the meteorologist Edward Lorenz in 1963, through a simple model of weather conditions that consisted of three coupled nonlinear equations. The model suggests "from two practically identical starting points, two trajectories would develop along for completely different paths, which made any long-term prediction impossible" (Lorenz 1662). The most popular example of the "Butterfly Effect" is the flapping of a butterfly's wings that shakes the air in Beijing and may cause, some time from now, a storm in New York. In this context, the concept of "Butterfly Effect" applies to different areas, including biology and astrophysics, where a small change, in iteration, in the initial state of a given system can result in either catastrophic or beneficial consequences, alter or annihilate (Gottlieb 2015). Another concept to understand the origin of nature's phenomena, biological systems, and our own history is fractal geometry. The fractal concept concerns the study of a wide variety of complex and irregular shapes and phenomena found in nature, presenting a repeated pattern found within a dynamic system (Mandelbrot 1982). Thus, any scale parts are in shape as the whole. This characteristic of fractals is called self-similarity (Figure 11.1). A classic example of self-similarity to illustrate this concept is given by Mandelbrot, "pulling out a piece of cauliflower, this little piece looks exactly like the whole cauliflower". This cauliflower could be separated into different pieces, looking as the initial cauliflower (Mandelbrot 1982). Thus, if the Big Bang theory is

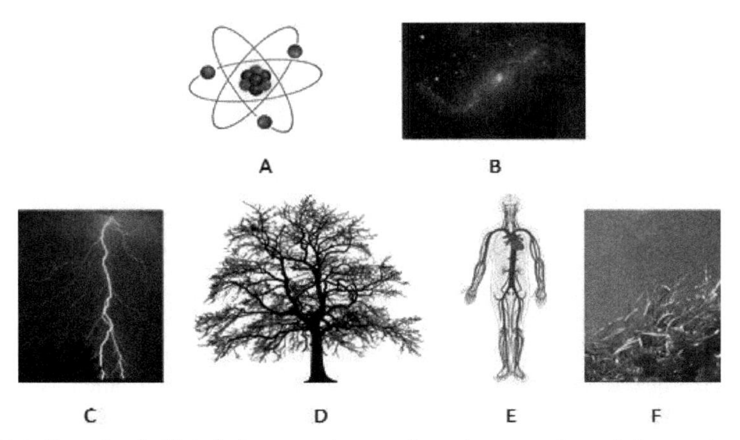

Figure 11.1. Example of self-similarity pattern between dynamic systems. Legend: A = atomic particles, B = galaxy (via lacteal), C = lightning, D = tree, E = human circulatory system, and F = seaweed. Source: https://pixabay.com/pt/.

true, everything that exists at this exact moment in history is a fractal of the starting point that caused the Big Bang (Figure 11.1).

Fractals also show two other striking features: infinite complexity (iteration) and fractal dimension. The first concerns the incessant repetition of a certain operation or geometric pattern. The second is a measure that quantifies the density of fractals in the metric space in which they are defined and serves to compare them, as well as it is related to its structure, its behavior, and its degree of irregularity, commonly expressed by complex numbers (Capra 1996, Gottlieb 2015). These are represented by points on the plane defined by the real and imaginary axes called the complex plane (Capra 1996). Complex numbers are described with the following equation:

$$z = x + iy$$

The complex variable is z, the real part is x, and the imaginary part is y. The fractal dimension is fractional, while the Euclidean dimension is entire. This mathematics of complexity has applications in all fields of science, including deep ecology and systemic thinking. In this case, we can extend this concept not only to the life of Earth's ecosystems, but also to the Big Bang. The theory of complexity and chaos added to the Butterfly Effect and fractal geometry may contribute to understanding the bases of its emergence. However, all philosophical, scientific, and technological knowledge is insufficient to combat and stop predatory actions of the *Homo sapiens* species in our common home: Planet Earth.

Human Evolution: Homo sapiens

From the original moment, the Big Bang, intelligence was necessary for architecture, physical engineering, chemical, biological, and energy; the matter with its different combinations, forms of life, all happened under dramatic conditions and intense natural selection, and a long time (billions of years) to culminate in the first primates, about 70–80 million years ago (Dalgalarrondo 2011, Futuyama 2009). Then, the first bipedal hominids that belong to our evolutionary lineage, appeared approximately between 6 and 7 million years ago. The species Homo sapiens appeared about 190,000 to 200,000 years ago (Dalgalarrondo 2011, Futuyama 2009). Regarding the evolutionary scale, modern human beings are extremely recent and with the immense destructive power of everything that originated them.

Throughout the 20th century, the oldest ancestor of the modern man known would be the genus *Homo erectus*, who lived in Africa around 4 million years ago. However, with the techniques of molecular biology, this scenario became more obscure. Biotechnology tools allow testing of not only the different models of modern human origins, but also the demographic history of exploration and the types of selection from different regions of the genome, and even specific characteristics have suffered (Disotell 2012, Kelso and Prüfer 2014, Lowery et al. 2013). Data suggests the origin of modern man comes from a gene mosaic of *H. erectus*, *H. heidelbergensis*, Neanderthal, and hominids from Siberia called Denisovans. However, sequencing analyses of additional hominid fossils are needed to clarify our origins and our history. Although there are controversies concerning our origins, the same is not true of the lifestyle of the ancient man. According to Palaeoarchaeology studies, Paleolithic men were nomads and hunter-gatherers. They did not grow or produce their own food, so energy wasting and infections were constant variables, as well as the high mortality rate and low life expectancy (Armelagos et al. 2005, Challa and Uppaluri 2019, Dyble et al. 2015, Macintosh et al. 2017, Vale 2015). Despite its low energy density, the Paleolithic diet was diverse, based on fruits, flowers, roots, tubers, eggs, seeds, nuts, crustaceans, insects, mollusks, and fish (Challa and Uppaluri 2019). Once guaranteed the survival of the species in the Paleolithic, humans faced new population challenges in the following period, the Neolithic (10,000 BC to 4,000 BC). In this period, the Agricultural Revolution was the crucial landmark of the old world and marked the dominance of man over food production and the establishment of sedentary life to the detriment of the nomad. The fast increase in population, lifestyle modification with animals' domestication, extensive damage of ecological balance (occupation of territories for the monoculture of food and insertion of exotic species), the beginning of non-pagan religions, and patriarchy were the trigger for the emergence of social inequality (Armelagos et al. 2005, Dyble et al. 2015, Vale 2015). The end of the Neolithic period was marked by extensive urbanization and exploration of new territories, resulting in intense epidemics and the concentration of power and wealth in the minority (Armelagos et al. 2005, Dyble et al. 2015, Vale 2015). Despite this, the human population overcame adversity, maintaining the reproductive and survival pattern, gradually achieving an increase in life expectancy. At the same time, there was a break in harmony between inter and intraspecific relations. An important data, might be directly related to the ecological crisis that our species created, the woman of the prehistoric period (Paleolithic and Neolithic) had much more access to human rights and freedom of expression (Vale 2015).

According to studies in the field of anthropology, investigations show men and women lived under an egalitarian regime, regardless of gender. According to Dyble et al. 2015, the division of tasks between men and women was very similar, including the care of children and children of the tribe (Dyble et al. 2015). Equality between genders may have been a great asset for the perpetuation of the species. The tribes were formed by a few individuals, and the efforts of all to obtain food, shelter, and security were the differential for their survival (Dyble et al. 2015).

Regarding sex life among members of the tribe, studies suggest prehistoric women had free will to decide who to relate to, even without genetic or marital ties.

Gender equality induced the most diverse social and marital arrangements. Even in groups of up to 20 people, the egalitarian model presented a 12% chance of not being related to another individual (no genetic or marital relationship), while those who lived in a non-egalitarian model presented less than 1% chance of not being related to any other individual (Dyble et al. 2015). Having unrelated people in a tribe or community was important to our evolution. Forming tribes of unrelated individuals, hunter-gatherers have developed the ability to cooperate with unrelated individuals (Dyble et al. 2015). However, this does not exclude the possibility of violence against prehistoric women. Archaeological and paleontological findings show women were stronger than paddle athletes, suggesting that in a situation of violence they had more defense using their physical strength than today's sedentary women (Macintosh et al. 2017). However, over time, inequality between genders increased and women began to occupy a more fragile role, in a situation of total dependence on a male figure, whether from the father, the brother, or the husband, originating the molds of a patriarchal and sexist culture.

At the end of the Neolithic demographic growth was increasing, as well as the patriarchal regime, the Judeo-Christian religions, the division of private property, tasks, powers, gender discrimination, and social inequality (Armelagos et al. 2005). Perhaps, this abrupt change in lifestyle was largely responsible for the disconnect between man and nature. As men developed the mentality for the creation of new realities, their animal part linked to nature was reduced, opening the path for ecological degradation. Ecological degradation must be understood not only as the destruction of the environment, but as a deterioration of the harmonic relations between the biotic and abiotic environment, where man is the main causative agent.

Human Ecology and Ecological Crisis

The term Ecology originates from the Greek "oikos", which means house, environment, family, race; in Greek it is formed from the verb *oikizein,* which means install, build, found. Logy was formed from the verb *leguein*: to say, to announce, to read, to order. The word logos (logy) is also attached to this verb, which means word, reason, discourse. Therefore, it would be the study of the house or the place where you live (Castro 1992).

For Odum (1988), ecology is the study of the structures and functions of nature, the structure and functions of ecosystems. Science is the study of interactions between organisms and their environment, and the totality of physical and biological factors that, directly or indirectly, affect and/or are affected by them (Odum 1988). Ecology is also an interdisciplinary and interactive study, that must synthesize information and knowledge from the majority or all other fields of knowledge (economic, social, political, cultural, and psychological, peculiar to man [anthropology]). Ecology is not the environment (Kormondy and Brown 2002). Human ecology studies the relationships between individuals and between different communities, as well as their interactions with the environment they live in, at an ecological, psychological, cultural, and social level. It describes how man adapts to the environment in different places on the planet, how food, shelter, and water are obtained. Therefore, the relationship with natural resources, air, water, fauna, and flora, as well as the relationships between individuals. Social issues such as rural exodus, uncontrolled growth in cities, urban infrastructure, as well as population characteristics (growth rate, density, birth and mortality rates, and average age) are addressed in this specialty. According to Castro (1992), human ecology is not primarily an economic and political problem, but rather a problem of the relationship of man with himself, with others, and with things. (...) Thus, when thinking about ecology beyond the environmental, we perceive something fundamental linked to the human essence, that "implies the meaning of man and the universe in his Being." From a biblical perspective, it is written in the genesis: "Be fruitful, multiply, fill the earth, and subdue it. Have dominion over the fish of the sea, over the birds of the sky, and over every living thing that moves on the earth." God said, "Look, I have given you every plant that produces seed on the face of the whole earth, and every tree that bears fruit that produces seed. It will be your food! (Editora Paulus 2002). This guideline transmits an authorization to man that nature must be dominated, subdued, and conquered for satisfaction (Bourdeau 2004). Some authors suggest the Western Judeo-Christian religion is the historical root of the socio-environmental and ecological crisis (Moncrief 1970, White 1967), since it probably influenced the idea of man's superiority over the rest of creation. Non-pagan religions would have led man to indifference towards the environment by offering support for the belief, not only that human beings are separated from nature, but also superior and with a divine mandate to dominate and explore it (Bourdeau 2004).

However, if there is a creator God or the Big Bang, all creation and not just humans were made according to the image and likeness of God or the Big Bang. Above all, the current Pope Francisco recognizes the biblical interpretation of the passage was not always so clear, highlighting the current exegesis allows to emphasize the call for protection and care of creation (Pope Francis 2015).

"If it is true that we, Christians, sometimes misinterpret the Scriptures, today we must reject the fact that since we are created as the image of God with the mandate to rule the earth, deduce an absolute dominion over all creatures. It is important to read the biblical texts in their context, with hermeneutics and remember we are invited to "cultivate and guard" the garden of the world" (according to Genesis 2,15) (Pope Francis 2015).

Thus, the responsibility for the crisis origin could not be credited to the Judeo-Christian tradition. Although it is not possible to state that its influence on Western thought did not contribute to the formation of an anthropocentric view of the universe, that relegates nature to the function of satisfying the needs of men (Moncrief 1970). Perhaps this feeling of human superiority over nature is an integral part of Being, sharpened by its mental, intellectual, scientific, technological, and physical capacity (bipedalism). Bipedalism and brain development resulted in great discoveries and advances for humanity, but also a hostile, aggressive, and thankless attitude towards nature. For a long time, it was also believed that natural resources were not finite, just as some still believe that science and technology can overcome the damage that we cause to nature. Or in the future we can inhabit other planets in the solar system. This way of thinking, investigating, and recreating nature is one of the legacies of René Descartes's thought (1596–1650) about the origin of the Universe (Descartes 2001). In the "Discourse on the Method", Descartes revealed his ideas on the main rules for scientific practice, which we have used until today. Descartes was one of the great classic rationalists of his time, laying the foundations for philosophical/scientific thinking. He broke with Aristotelian tradition and scholastic thought, dominating philosophy in the medieval period. The separation between subject and object of knowledge has become fundamental for all modern philosophy, resulting in the so-called reductionism mechanism or Cartesianism (Descartes 2001). Except for the errors and limitations of his philosophical/scientific thinking, a great contribution from Descartes was the rational elaboration of a practical method to test hypotheses, a philosophical reflection for a scientific and medical practice that leads to the truth. However, regarding nature, ecosystems, and life itself, when separating the subject from the object for investigative purposes, the chance of making mistakes or obtaining false results is a reality and a risk. Therefore, systemic thinking and deep ecology obtained space in the deconstruction of the mechanistic, reductionist, and anthropocentric vision for the construction of a new paradigm of existence. Thus, despite being just another specie within this enormous biodiversity, man is endowed with intelligence and awareness, responsible for ensuring the healthy perpetuation of the planet.

Deep Ecology, Systemic Thinking, and Bioethics

This paradigm arose to counter the term "shallow ecology", the dominant model for the indiscriminate use of natural resources, where humans are the center of everything, and nature is something to be explored (Naess 2007, 2008). Deep ecology understands nature and the world not as a collection of isolated objects, but as a network of phenomena fundamentally interconnected and interdependent. Deep ecology recognizes the intrinsic value of all living beings and conceives human beings only as a thread in the web of life (Capra 1996, Naess 2007, 2008). Systemic thinking also follows the same philosophical line of deep ecology, since it diverges from mechanistic thinking (advocates the separation of parts of study object for analysis, investigation, and understanding), proposing that systematic understanding means putting in context, establishing the nature of relationships or their interdependence. However, systemic thinking requires a comprehensive and profound knowledge of natural phenomena, which is humanly impossible (Capra 1996, Naess 2007, 2008). However, the Cartesian model is based on the certainty of scientific knowledge, but science is often not able to offer complete and definitive explanations or solutions to global ecosystem problems, since it isolates man from ecological context. In the current socio-environmental scenario, men at the center of the discussions are essential to finding the best actions and solutions for the ecological crisis created by them. However, being at the center should not mean an anthropocentric stance, but rather a democratic ecocentrism, in which the rights of nature to follow its designs are guaranteed and respected, and our commitment to fulfill duties to safeguard the ecological relations of planet Earth is maintained, according to bioethical principles (Capra 1996, Naess 2007, 2008).

The movement in defense of animal and plant rights is not new, starting in 1927 in Germany, with the pastor from Halle an der Saale Fritz Jahr (Fritz 2012). In 1927, he critically opposed Kant's categorical imperative with his bioethical imperative: "Respect every living being essentially such as an end in itself and treat it as such if possible!" (Muzur and Sass 2012). Then, Potter (Potter 1971) would bring the original vision of the global commitment to the balance and preservation of the relationship between human beings and the ecosystem, and the life of the planet with a new definition:

"I propose the term Bioethics as a way of emphasizing the two most important components for achieving new wisdom, which is so desperately needed: biological knowledge and human values" (Potter 1971).

It is a segment of ethics (philosophy of morals). One of the concepts that define bioethics (ethics of life) is the science "which aims to indicate the limits and the purposes of man's intervention in life, to identify the rationally possible reference values, to denounce the risks of possible applications" (Leone et al. 2001). Its purpose is not to develop new general ethical principles, but to apply these principles to the scope of life sciences and health care, especially to the new emerging problems (Ferrer and Alvarez 2003, Levine 2007). Bioethical principles have been proposed (Belmont Report 1979) (National Commission for the Protection of Human Subjects of Biomedical and Behavioral Research 1979) to guide research with human beings, totally focused on the value of human life. In 1979, Beauchamp and Childress (1979) in the work Principles of biomedical ethics, extended the use of these principles of medical practice to all those concerned with people's health. This means humanistic bioethics, in which the life and health of Planet Earth, our *common Oikos,* was relegated to remain without due care and attention, although it must serve our wills. Therefore, beneficence, non-maleficence, autonomy, and justice are *prima facie* principles that must be extended to all nature. Bioethics, combined with systemic thinking and deep ecology, form the triad capable of promoting new knowledge, behaviors, and awareness of ecological chaos that humans insist on infringing on nature and, lately, themselves. Thus, "*One Health*" can change the hegemonic model of man's domination over nature, through its integrative and interdisciplinary approach to knowledge to prevent global epidemiological outbreaks.

One Health for Our Common Oikos

The concept of *One Health* proposes to integrate human, animal, and environment health for the prediction and control of diseases in the human-animal-ecosystem interface (American Veterinary Medical Association 2008, Coker et al. 2011, Food and Agriculture Organization, OIE and World Health Organization 2008, Kakkar and Abbas 2011). I is also defined as collaborative efforts across multiple disciplines, working locally, nationally, and globally to achieve optimal health for people, animals, and the environment (American Veterinary Medical Association 2008, Coker et al. 2011, Food and Agriculture Organization, OIE and World Health Organization 2008, Kakkar and Abbas 2011). This approach recognizes the need for public, animal, and environmental health leaders to develop research, create evidence, and guide public policy decision-makers on collaboration between sectors, thus aiming to reduce the emergence and resurgence of zoonotic diseases. Therefore, *One Health* is becoming a goal of contemporary global health. However, the use of the concept is not restricted to zoonoses. Its possible application is broader, mainly regarding food, environment, and sustainable development. Health management requires a holistic *One Health* perspective, recognizing the complex interactions between human health, livestock, pet and wildlife health, climate, ecosystems, agriculture, food production systems, and human development (Food and Agriculture Organization, OIE and World Health Organization 2008, Kakkar and Abbas 2011). This concept has been proposed to protect human health from outbreaks, epidemics, and pandemics from infectious diseases. Although, it is an excellent opportunity to attract the attention of civil society and world governments to the issues of habitat devastation of natural hosts and vectors of human diseases. Destruction, reduction, contamination, restriction of the biodiversity of ecosystems and natural habitats by humans can be considered one of the causal factors for the worldwide epidemiological outbreaks. Only a systemic, transdisciplinary, and integrative approach can ensure sustainable ecological and bioethical management of global health, at a time in the history of total depletion of natural resources, climate change, social, racial and gender inequalities, and poverty. We are a century and a half late in changing our *modus operandi* of existence, causing irreparable consequences. Therefore, the current paradigm of perceiving nature as an inanimate object, usurping it, manipulating it, exploiting it, destroying it, contaminating it, for indiscriminate profit and consumption, urgently needs to be changed for an ecosystem friendly model, being bioethical and self-sustainable is urgent. The paradigm shift includes a change in social organization, from hierarchical models to networks. The conversion of post-industrial society to this new paradigm is essential. It requires a reinterpretation of the concept of Progress, contemplating greater harmony and systemic balance between the whole and the parts (Capra 1996), promoting quality, and not just the quantity of growth and development. For reconnection and respect for our *common Oikos,* priority actions first proposed in 1972 are necessary to be followed (United Nations Environment Program 1972). After this first meeting, where 113 countries participated, other meetings occurred to discuss, propose actions, define voluntary goals and commitments of the countries involved. Thus, regularly monitoring the progress of countries in their socio-environmental goals, reducing global warming and sustainable development were discussed. Other important examples were ECO-92, which produced Agenda 21 and COP21. Agenda 21 was a document that established the importance of each country to commit itself to reflect, globally and locally, on how governments, companies, non-governmental organizations, and all sectors of society could cooperate in the study of solutions for socio-environmental problems (Brasil 2004). Each country developed its Agenda 21. In Brazil, the National Agenda 21 established the following priority actions (Commission on Sustainable Development Policies and the National Agenda 21 2004):

- **Social inclusion:** access of the entire population to **Education, Health, and Income Distribution**;
- Urban and rural sustainability, preservation of natural and mineral resources and **Political ethics** for planning towards **Sustainable Development**;
- And, the most important of these priority actions according to the study, the planning of **Sustainable Production and Consumption Systems** against the culture of waste.

Concern on global environmental issues reached its peak at the end of the 20th century with discussions around climate change. The Kyoto Protocol (Meneguello and Castro 2007, Moreira and Giometti 2008, Souza and Corazza 2017, United Nations 2016, 1998) was signed in 1997, due to the 3rd Conference of the Parties to the Convention on Climate Change. Industrialized countries committed to reduce 5.2% by 2012 their emissions of gases that contribute to global warming, calculated based on 1990 emission levels. At the time, the USA, Iraq, Afghanistan, among other countries, did not sign a commitment to reduce gas emissions. Canada, Russia, and Japan, which have ratified the Protocol, also did not agree to sign new reformulated commitment period if the Americans and major emerging economies, such as China, did not do their part (Souza and Corazza 2017, United Nations 2016). In 2012 the protocol expired, and studies demonstrated that greenhouse gases increased by 16.2% between 2005–2012. However, 37 European Union countries exceeded 5% reduction (AEA 2015, CAIT 2016). The 20th edition of the Conference of the Parties to the United Nations Framework Convention on Climate Change, COP 20, was performed in December 2014 in Lima, Peru (IISD 2014). During the event, the Lima Call for Action on Climate was prepared, a document with the basic elements for the new global agreement that would replace Kyoto, approved during COP 21. At COP 21, a new global agreement was approved, the Paris Agreement, with targets for reducing greenhouse gas emissions for all countries, developed and developing, defined nationally according to the priorities and possibilities of each country (Afionis 2017, Boucher et al. 2016, Dimitrov 2016, Esteves et al. 2016). The European Union (EU) released its proposal to reduce greenhouse gases for the new global treaty. The countries indicated the Intended Nationally Determined Contributions (INDC) to reduce emissions by at least 40%, by 2030, compared to 1990 levels. The new agreement will come into effect from 2021 (Afionis 2017, Boucher et al. 2016, Dimitrov 2016, Esteves et al. 2016). Several conventions have been happening since the 1970s, on biodiversity, climate, and climate change. Although countries are committed to complying with the agreements, little has been effected for the preservation and respect of nature, to guarantee universal access to drinking water, quality atmospheric air, food, housing, decent employment, and poverty reduction. Why do agreements fail? This is a great debate question to solve the heart of the problem. Many external factors can be listed. However, the time is right to discuss the philosophical, biological, and emotional causes to find a plausible answer to the usurper and disrespectful behavior of man towards nature. Perhaps the first dominant factor is the certainty of finitude by the individual, behaving as a true parasite of nature. Use the most natural resources in this life, as you do not know if you will have another opportunity (life after death). There is no certainty when natural resources will end, so man dominates, usurps, exploits, and accumulates as much as possible to leave an inheritance for descendants. Such behavior implies two major mistakes. The first is characterized by leaving a terrible example to be perpetuated by descendants. The second concerns a very peculiar human characteristic: loving and caring more for those who share the same moral, social, and religious code. Loving your neighbor may imply restricting the circle of relationships. There may be difficulty to welcome the different, who do not have the same language, who seem to be inanimate, and who do not share the same codes of ethics and religious practices. The background of this scenario of a deep scarcity of emotional intelligence is the lack of self-knowledge, self-esteem, and systemic thinking, basic concepts of philosophy, physics, chemistry, mathematics, biology, and mainly compassion. In this context, compassion could be understood as a feeling when you witness the illness (suffering, stress) of the other (or yourself) and it generates motivation or desire to help (or to take care). Thus, promoting feelings of belonging, happiness, security, love, greater quality of life, and mutual health. The moment we perceive the other as ourselves, the feeling of compassion comes automatically. However, if we confuse compassion with pity, a feeling of superiority may emerge, which may trigger conflicts, disputes, and harassment. Loving what is different causes fear and insecurity, which is biologically natural. Although, we must transcend the biological, in the sense of survival, to be able to consolidate one health for our *common Oikos*. In this sense, education plays an important role in transcending the purely biological, material, finite, and segmented to a feeling of compassion, respect, allocentrism, gratitude, surrender, magnanimity, permanence, and belonging. The great paradox is that to transcend the biological, it is necessary to know it deeply, since self-knowledge is fundamental. Currently, self-knowledge is widely discussed, but in the sense of identifying patterns of thought and personal habits and, from that, allowing the individual to improve behavioral responses and decision making. However, for this ability to be developed, individuals need to know their biology. What matter, molecules, and cells the body is made of. How DNA, RNA, cells, and organs work. How energy is formed within cells, what nutrients are needed to maintain the body health, hours of sleep and, most importantly: knowing that all beings that inhabit planet Earth have the same origin and are formed from the same atoms and particles. What differentiates us is the expression of its form or manifestation in nature, according to physical, chemical, and biological laws (genetic-evolutionary). In this sense, the scientific, systemic, and secular study of the origin of the Universe, of life and biology, compared in schools since the process of literacy, can immensely contribute to the solid construction of this new paradigm of *One Health* for our *common Oikos*.

Urgent Priority Actions for Building an Ecologically Healthy Era for the Planet

The alternatives thought, discussed, and agreed between different governments of the world for reducing the impact of humanity on Earth is not only their responsibility, but ours as well. Our daily attitudes, such as consumption pattern, the disposal culture, housing models, highly polluting energy matrix, the means of transport we use, the way we dispose of our garbage, the time we spend in the bath, the modes of production, excesses and food waste, among many other attitudes, directly reflect on the environment and consequently on our life, our health, and on the planet. Therefore, regardless of how our politicians and government are facing and launching guidelines for the preservation and mitigation of damage caused to nature, civil society must start individually planning to build an ecologically healthy lifestyle. Self-knowledge, innovation, and entrepreneurship are the keywords of today and can be applied in our day-to-day for us to live in accordance with nature. However, breaking the *modus operandi* is important, imported from developed western countries and passed on from generation to generation. Perhaps, the first label to be modified is that human beings are endowed with a superior brain and better than other beings of nature. Such understanding generates discrimination, exclusion, hierarchies, low self-esteem, neglect, and disinterest in the collective and the other inhabitants of the Earth. Developing the awareness that no type of brain is superior to another in nature is fundamental. Moreover, there are different types of intelligence in humans, and each individual must identify aptitudes and limitations to improve or supply it. The second label concerns the progress and development of countries, where we still insist on the predatory model of civilization. Thus, nature is only a source of wealth, which must be exploited and exported to the maximum to generate profits for a minority of society. Meanwhile, most of the society (the poor) becomes hostage to a perverse and supremacy system that generates poor income distribution and all types of inequalities. Without preserved rights to a dignified life, such as the right to land and housing, food, public security, health, quality education, and freedom, how can these individuals be concerned with preserving the environment, if their capacity for self-preservation is precarious? This needs to be changed. A third imaginary also needs to be changed and our culture is deeply rooted, causing neuropsychiatric suffering (depression, anxiety, anguish, low self-esteem, and suicide) and physical (obesity and sedentary lifestyle), especially in young people: the successful adult has body patterns dictated by fashion (cult of the body at the expense of health), at least higher education and postgraduate education, stable employment with an excellent salary, house, electronics, cars, superfluous consumer goods that will be quickly disposed of (planned obsolescence), wife, children (who study in private schools, take a language course, have their tablet and cell phone, and practice some sport), health insurance and enough money for local trips and abroad. Biologically speaking, being successful is only about having a healthy physical and mental condition to leave descendants and reach the average life expectancy for the species. As a species, we are not just biology and we cannot ignore the cultural, scientific, and technological evolution that permeates us, often dictating the norms of living in society. Although, we need to properly use the resources offered by nature and knowledge to know the most ecologically healthy choices. Some simple measures can be adopted to build *One Health* for our *common Oikos:*

Governments

- Put into practice the fulfillment of all agreements established (between organizations, governments, and civil society) for the preservation of biodiversity, reducing greenhouse gas that contributes to global warming, application of Clean Development Mechanisms, through financing projects that contribute to the reduction of emissions, or that compensate them for the sequestration of polluting gases in the atmosphere;
- Make the United Nations Environment Program (UNEP) an oversight agency or government proposals and actions for the preservation of global natural resources. An agency that represents the interests of society through investigation of facts that characterize environmental crime, protection of victims and witnesses, and ownership and support of public criminal action is essential. However, it should have autonomy to perform its actions, regardless of the powers of the State. It is an agency that does not respond to any of the classic powers, as a subordinate hierarchically and without its assets;
- Invest heavily in replacing the fossil fuel-based energy matrix with clean energy (wind, solar);
- Foster projects for higher-level academics to promote debates on global issues (biodiversity and ecosystems, environment, genetic heritage, sustainable socio-

economic development, ecologically sustainable food production, terrorism, biopiracy, clean energy, exploitation of natural resources, contemporary slavery, eradication of corruption and poverty, and epidemiological transition) in public and private schools, along the lines of General Assembly of the United Nations;

- Invest in scholarships for high school students to develop projects for ecologically intelligent solutions to the socio-environmental problems of their cities or regions;
- Invest in self-organized and self-sustaining cities and neighborhoods, with ecological housing, vegetable gardens, and community orchards, with employment and income generated in these spaces, sewage treatment, and collection of organic matter for energy generation by biodigesters;
- Supervise the production of the chemical industry, the commercialization and use of products in agriculture;
- Finance scientific projects in Public and Private Universities for the decontamination of water, soil, and air;
- Establish a percentage of the country's Gross Domestic Product to invest in these actions; and failure to comply with the agreements should result in economic sanctions to the country.

Civil Society

- Request governments to include environmental education and deep ecology from the basis of education (daycare centers, elementary and high school);
- Create associations, councils, and management committees to supervise the actions and the fulfillment of the goals established by the governments;
- Create groups on the web to inform and discuss socio-environmental problems in your neighborhood;
- Create groups of meditation and philosophical-religious discussions;
- Guide families to encourage children to meditate and practice joint activities with family members, preventing children from keeping busy with electronic devices and virtual media, consuming ultra-processed food to occupy time;
- Adopt a healthy lifestyle with a diet based on natural products, avoiding industrialized products and red meat;
- Participate more in the neighborhood life through the occupation of public spaces (squares), walks, bike rides, conversation circles, games, and sports activities;
- Take care and preserve the good conditions of public spaces in your neighborhood;
- Report to the competent agencies the acts of depredation and vandalism of public spaces, as well as the disposal of solid and organic residues in inappropriate places;
- Consume less packaged products, as they generate solid waste;
- Separate and dispose of waste correctly;
- Make compost with organic waste;

- Build cisterns in homes to collect rainwater for use in the bath and toilet, washing dishes and clothes, watering plants, etc.
- Require government funding to place solar collectors in homes;
- Require your neighborhood to have product fairs with organic certifications;
- Engage the neighborhood to create the community vegetable garden and orchard;
- Do not agree with deforestation, grounding, and chemical weeding

Conclusion

Homo sapiens is a recent species on Earth and exists due to the appearance of the first photosynthetic organisms, allowing oxygen to be accumulated in our atmosphere. Thus, after the evolutionary succession, it presents a dissociative and predatory behavior with the nature that originated it. The creature turns against the creator. We are living in a critical moment that affects us deeply. For some reason, we are not aware that destroying, exploring, polluting, and wasting the natural resources the planet offers us for free, without demanding anything in return, we put ourselves in danger of extinction. History, geography, and biology show us that even after several cataclysms and mass extinctions nature itself moves, in some way still unknown, and life is reborn by itself. Life or nature itself is the source and origin of everything, therefore self-creative, self-organized, and autopoietic. The beings come from it and not the opposite. However, we are contributing to accelerating the mass extinction process on Earth. It is crucial to immediately change the *modus operandi* in government systems, in terms of income generation, production, development, education, laws, human behavior and thinking, thus leaving a habitable planet for the next generations. Deep ecology, systemic thinking, and the concept of One Health can contribute to this new paradigm of existence. Now is the time to build new relationships with the inhabitants of the planet, and to understand that nature has an intrinsic value and not just instrumental. Self-knowledge, compassion, building a culture of encounter, care, respect for others, mineral, vegetable, or animal, are powerful tools for the preservation of our common *Oikos* and for the conversion to our common essence: nature...or better, our habitation.

Acknowledgment

I would like to thank my husband Guilherme Dornelles, an environmentalist, member of local, regional, and national health and environment councils, for his in-depth knowledge of health and environment issues, and for being the source of my inspiration.

References

AEA - Agência Europeia do Ambiente. (2015). O Ambiente na Europa: Estado e Perspetivas 2015 – Relatório síntese. Agência Europeia do Ambiente.

Afionis, S. (2017). The European Union in International Climate Change Negotiations. University of Leeds, UK.

Armelagos, G.J., Brown, P.J. and Turner, B. (2005). Evolutionary, historical and political economic perspectives on health and disease. Soc Sci Med 61: 755–65.

AVMA – American Veterinary Medical Association. (2008). One Health: A New Professional Imperative. One Health Initiative Task Force: Final Report.

Beauchamp, T.L. and Childress, J.F. (1979). Principles of Biomedical Ethics. Oxford University Press, New York.

Boucher, O., Bellassen, V., Benveniste, H., Ciais, P., Criqui, P., Guivarch, C. et al. (2016). Opinion: In the wake of Paris Agreement, scientists must embrace new directions for climate change research. Proceedings of the National Academy of Sciences 113: 7287.

Bourdeau, P. (2004). The man-nature relationship and environmental ethics. J Environ Radioact 72: 9–15.

Brasil, Ministério do Meio Ambiente. (2004). Agenda 21 Brasileira. Decreto Presidencial de 03 de fevereiro de 2004.

CAIT – Climate Data Explorer. (2016). World Resources Institute, Washington, DC.

Capra, F. (1996). A teia da vida. Uma nova compreensão científica dos sistemas vivos. Editora Cultrix, São Paulo.

Castro, M.A. (1992). Ecologia: a cultura como habitação. pp 13–33. *In*: Soares, A. (ed.). Ecologia e literatura. Tempo Brasileiro, Rio de Janeiro (RJ).

Challa, H.J. and Uppaluri, K.R. (2019). Paleolithic Diet. StatPearls [Internet]. StatPearls Publishing, Treasure Island (FL).

Coker, R., Rushton, J., Mounier-Jack, S., Karimuribo, E., Lutumba, P., Kambarage, D. et al. (2011). Towards a conceptual framework to support one-health research for policy on emerging zoonoses. Lancet Infect Dis 11: 326–31.

CPDS – Comissão de Políticas de Desenvolvimento Sustentável e da Agenda 21 Nacional. (2004). Agenda 21 Brasileira: Ações Prioritárias. CPDS, Brasil.

Dalgalarrondo, Paulo. (2011). Evolução do Cérebro: Sistema Nervoso, Psicologia e Psicopatologia Sob a Perspectiva Evolucionista. ArtMed, Porto Alegre (RS).

Descartes, R. (2001). Discurso do método. Martins Fontes, São Paulo.

Dimitrov, R.S. (2016). The Paris agreement on climate change: Behind closed doors. Glob Environ Polit 16: 1–11.

Disotell, T.R. (2012). Archaic Human Genomics. Am J Phys Anthropol 55: 24–39.

Dyble, M., Salali, G.D., Chaudhary, N., Page, A., Smith, D., Thompson, J. et al. (2015). Sex equality can explain the unique social structure of hunter-gather. Science 348: 796–798.

Editora Paulus. (2002). Bíblia de Jerusalém. Paulus Editora, Porto Alegre (RS).

Esteves, P., Santos, M. and Amorim, A. (2016). Resultados da COP 21 e a participação do BASIC BPC. Policy Brief.

Food and Agriculture Organization, OIE and World Health Organization. (2008). Contributing to One World, One Health. A Strategic Framework for Reducing Risks of Infectious Diseases at the Animal-Human-Ecosystems Interface. Report.

Ferrer, J.J. and Alvarez, J.J. (2003). Para fundamentar a bioética. Loyola, São Paulo (SP).

Fritz, J. (2012). Bioética 1927: Revendo as relações éticas dos seres humanos com animais e plantas. pp. 439–442. *In*: Pessini, L., Barchifontaine, C.P., Hossne, W.S. and Anjos, M.F. (ed.). Ética e bioética clínica no pluralismo e diversidade. Teorias, experiências e perspectivas. Centro Universitário São Camilo/Ideias & Letras, São Paulo (SP).

Futuyama, D. (2009). Biologia Evolutiva: Editora Funpec, Ribeirão Preto (SP).

Gamow, G. (1948). The evolution of the Universe. Nature 162: 680–682.

Gottlieb, M.G. (2015). From the Big Bang theory to the fractal geometry: a systemic reflection for Alzheimer's disease. PAJAR 3: 22–28.

Holm, S. (2002). Does chaos theory have major implications for philosophy of medicine? Med Humanit Rev 28: 78–81.

IISD Reporting Services. (2014). Earth Negotiation Bulletin. A Reporting Service for Environment and Development Negotiations. Summary of the twelfth session of the UN general assembly open working group on sustainable development goals: 16–20 June 2014 Vol. 32. New York.

Kakkar, M. and Abbas, S.S. (2011). One health: moving from concept to reality. Lancet Infect Dis 11: 808.

Kelso, J. and Prüfer, K. (2014). Ancient humans and the origin of modern humans. Curr Opin Genet Dev 29: 133–138.

Kormondy, E.J. and Brown, D.E. (2002). Ecologia Humana. Atheneu, Rio de Janeiro (RJ).
Leone, S., Privitera, S. and Cunha, J.T. (2001). Dicionário de Bioética. Editorial Perpétuo Socorro/ Santuário, Aparecida (SP).
Levine, C. (2007). Analyzing pandora's box: The history of bioethics. pp. 3–23. *In*: Eckenwiler, L.A. and Cohn, F. (eds.). The Ethics of Bioethics: Mapping the Moral Landscape. Johns Hopkins University Press. Baltimore.
Lorenz, E.N. (1662). Deterministic nonperiodic flow. J Atmos Sci 20: 130–141.
Lowery, R.K., Uribe, G., Jimenez, E.B., Weiss, M.A., Herrera, K.J., Regueiro, M. et al. (2013). Neanderthal and Denisova genetic affinities with contemporary humans: introgression versus common ancestral polymorphisms. Gene 530: 83–94.
Macintosh, A.A., Pinhasi, R. and Stock, J.T. (2017). Prehistoric women's manual labor exceeded that of athletes through the first 5500 years of farming in Central Europe. Sci Adv 3: eaao3893.
Mandelbrot, B.B. (1982). The fractal geometry of nature. W.H. Freeman and Company, New York.
Meneguello, L.A. and de Castro, M.C.A.A. (2007). O Protocolo de Kyoto e a geração de energia elétrica pela biomassa da cana-de-açúcar como mecanismo de desenvolvimento limpo. Interações 8: 33–43.
Moncrief, L.W. (1970). The cultural basis for our environmental crisis. Science 170: 508.
Moreira, H.M. and Giometti, A.B.R. (2008). Protocolo de Quioto e as possibilidades de inserção do Brasil no Mecanismo de Desenvolvimento Limpo por meio de projetos em energia limpa. Contex Intern 30: 9–47.
Muzur, A. and Sass, H.M. (2012). Fritz Jahr and the foundations of global bioethics: the future of integrative bioethics. Lit Verlag, Hamburg, Germany.
Naess, A. (2007). Los movimientos de la ecología superficial y la cología profunda: un resumen. Rev Amb Desarrol Cip 23: 102–105.
Naess, A. (2008). Life's Philosophy. Reason & Feeling in a Deeper World. The University of Georgia Press, Athens and London.
National Commission for the Protection of Human Subjects of Biomedical and Behavioral Research. (1979). The Belmont Report. Ethical Principles and Guidelines for the Protection of Human Subjects of Research. Department of Health, Education, and Welfare. Government Printing Office, Washington, DC.
Odum, E.P. (1988). Ecologia. Guanabara Koogan, Rio de Janeiro (RJ).
Oestreicher, C. (2007). A history of chaos theory. Dialogues Clin Neurosci 9: 279–289.
Potter, V.R. (1971). Bioethics. Bridge to the future. Prentice Hall, Englewood Cliffs.
Santo Padre Francisco. (2015). Carta Encíclica Laudato Sí: sobre o cuidado da casa comum. Paulinas, São Paulo (SP).
Sharma, V. (2009). Deterministic chaos and fractal complexity in the dynamics of cardiovascular behavior: perspectives on a new frontier. Open Cardiovasc Med J 3: 110–123.
Souza, M.C.O. and Corazza, R.I. (2017). Do Protocolo Kyoto ao Acordo de Paris: uma análise das mudanças no regime climático global a partir do estudo da evolução de perfis de emissões de gases de efeito estufa. Desenvol Meio Amb 42: 52–80.
UNEP – United Nations Environment Programme. (1972). Stockholm Declaration on the Human Environment. Report of the United Nations Conference on the Human Environment.
United Nations. (1998). Kyoto Protocol to the United Nations Framework Convention on Climate Change.
United Nations. (2016). Framework Convention on Climate Change (UNFCCC). The Paris Agreement.
Vale, A. (2015). A mulher e a Pré-História: alguns apontamentos para questionar a tradição e a tradução da mulher-mãe e mulher-deusa na Arqueologia pré-histórica. Conimbriga 54.
White, L. (1967). The historical roots of our ecologic crisis. Science 155: 1203.

Index

N

O

P

S

T

U

9780367682750